KNOWLEDGE WITHIN GROUPS – FORMATION OF COMMUNITIES

AND NETWORKS

सिद्धिमूलं प्रबन्धनम्
भा. प्र. सं. इन्दौर
IIM INDORE

A Doctoral Dissertation Submitted in Partial Fulfilment of the Requirements for the Fellow

Program in Management Industry)

Indian Institute of Management Indore

By

Vinay Avasthi

2018

Thesis Advisory Committee:

Prof. Shubhamoy Dey
(Chairperson)

Prof. Kamal K. Jain | Prof. Rajhans Mishra | Prof. Shankar Venkatagiri
(Member) | (Member) | (External Member)

ABSTRACT

Today's modern enterprises are entities that strive to create, share, and consume knowledge effectively with the least amount of overhead possible. The endeavor is to enable business performance using all the resources that are available to it. Businesses organize themselves into different groups, each performing specific tasks or projects, and each member defined by the roles and responsibilities of its management. Individuals and groups clearly make use of knowledge, both explicit and tacit, in what they do.

Communities of practice, or CoPs, are custodians of knowledge that an enterprise aspires to, but which they do not directly create. Individuals within the communities create the knowledge; it is the job of communities to integrate and institutionalize it. Even though organizations think of knowledge in organization, groups, and CoPs, it is necessary to focus in this study on individuals in the context of knowledge creation. Enterprises roll out knowledge management initiatives for these purposes but have met with limited success because individuals are incentivized to use the knowledge to achieve business goals; there is no (or very little) incentive to share or seek knowledge.

Most of the knowledge management systems have an emphasis on the codification of knowledge. Codified knowledge is important, but real productivity enhancers in groups and organizations are sources of tacit knowledge that exist within the groups and individuals within those groups.

In this study, the knowledge within CoPs, groups, and individuals, is examined. Primarily, a knowledge ontology is defined that is appropriate in the context of CoPs, groups, and individuals while keeping in mind creation, codification, and usage of knowledge by individuals within the context of CoPs and groups.

It is established through quantitative analysis that organizations and their members miss the knowledge which goes along with employees who leave an organization. Enterprises cannot codify all the knowledge that they desire. One of the ways to access the tacit knowledge of an individual is socialization. A mechanism is defined to build knowledge networks based on the communication patterns across individuals as part of their interaction in a CoP. These knowledge networks are then examined using social network analysis reveals the fact that each group has leaders, matchmakers who hold on to their position for long periods based on their contribution. These results are valid across multiple domains.

Enterprises may have their taxonomies it makes the content unsuitable for natural language processing. Even synonyms for words may be unsuitable. We define the concept of knowledge adjacency to find more appropriate alternative phrases for a candidate query that user may want to search. The information communication system is sometimes used by the organization members to advance their aims giving rise the phenomenon of cliques. We look at communities of practice for the existence of the similar phenomenon. Additionally, a methodology is proposed to identify these cliques and evaluate their characteristics.

We also look at what is holding these communities of practice alive. We define a mechanism to identify these individuals whose attrition may be most damaging to the community.

Finally, we propose a system architecture that can be the foundation for next generation of knowledge management systems which is more helpful in identifying the knowledge and source of knowledge within a community of practice.

Keywords: Communities of Practice, Groups, Knowledge, Tacit Knowledge, Knowledge Management, Explicit Knowledge, Codified

ACKNOWLEDGEMENT

Firstly, I would like to express my sincere gratitude to the chairperson of thesis advisory committee Prof. Shubhamoy Dey for the continuous support of my research and related study, for his optimism, patience, motivation, and immense knowledge. His guidance helped me in all the time of research and writing of this thesis. I could not have imagined having a better advisor and mentor for the dissertation.

Besides chairperson of the committee, I would like to thank the rest of thesis committee: Prof. Kamal K. Jain, Prof. Rajhans Mishra, and, Prof. Shankar Venkatagiri, for their insightful comments, encouragement. Their hard questions motivated me to incorporate all the perspectives in the research.

A very special gratitude goes out to all the staff in FPM program office for providing extremely valuable support during this program.

I am grateful to my family for providing the support during this program.

And finally, last but not the least, my gratitude to everybody in my batch for continuing to encourage during this period.

TABLE OF CONTENTS

LIST OF FIGURES

ix

CHAPTER 1

1. Introduction

Knowledge workers think for a living; They live by their wits – any heavy lifting on the job is intellectual and not physical. They solve problems, understand and meet the needs of customers. They make decisions and collaborate and communicate with other people in the course of doing their work. Firms with the highest degree and quality of knowledge workers tend to be the fastest-growing and most profitable. Knowledge workers tend to be closely aligned with the organization's growth prospects. (Davenport, 2005)

Organizations and groups use knowledge on a regular basis. There are two types of knowledge: explicit knowledge and tacit knowledge. Explicit knowledge is what can be embodied in a code or a language and, as a consequence, it can be verbalized and communicated, processed, transmitted and, stored relatively easily. In contrast, tacit knowledge is personal and difficult to formalize – it is rooted in action, procedures, commitment, values, and emotions, etc. Tacit knowledge is the less familiar, unconventional form of knowledge. It is the knowledge of which people are not conscious. Tacit knowledge is neither codified nor communicated in a language; it is acquired by sharing experiences, by observation and imitation(Kikoski & Kikoski, 2004).

Organizations also tap into CoPs to get access to knowledge that they need. CoPs are informal with their organization, agendas, and leadership. (E. Wenger, 1998).

1.1 Modern enterprise: a knowledge-centric entity

The modern enterprise needs to have a workforce that can perform a variety of tasks. There is a significant shift in enterprises moving from a skill-based organization to a role-based organization. The role-based organization is more resilient and does not become outdated as early as the skill-based organization. Figure 1 illustrates a view of modern enterprise where the knowledge drives every facet of the enterprise. Business processes depend on roles, policies, organization structure, computing infrastructure, human resources, and policies. Each of these aspects is influenced by the role of knowledge within the organization (Burlton, 2001).

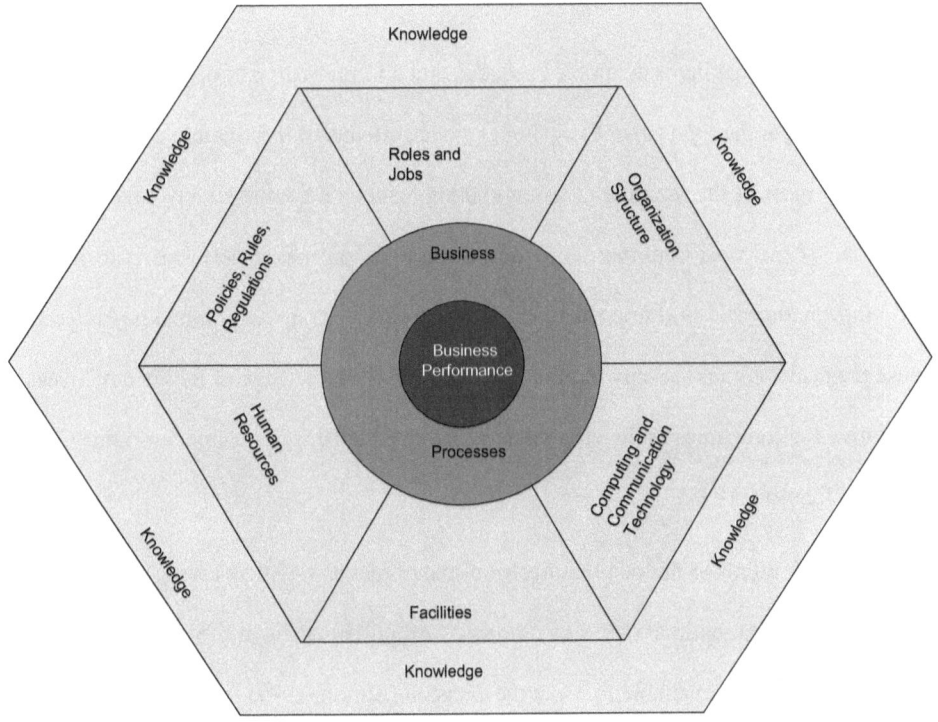

Figure 1 Modern enterprise, a knowledge-centric entity (adapted from (Burlton, 2001))

1.1.1 Role-based organization

The modern enterprise is becoming larger and geographically distributed. The structure of companies is also becoming more complicated, thus necessitating people in the enterprise to have access to other individuals based on their roles and expertise. People need to communicate with others based on the role they play in the organization (Stough, Eom, & Buckenmyer, 2000).

The concept of a role-based organization is not new. However, in the past organizations have found it difficult to implement this because of a lack of support from computing and communication technologies. Today, mobile computing and cloud computing are becoming common and affordable, and organizations are finding it easier to make use of the competencies

13

of their employees by structuring themselves as role-based virtual organizations (Foster, Kesselman, & Tuecke, 2001).

1.1.2 Just enough information, just in time

Enterprise decision-making is a complex process, and an important part of this is access to the right information at the right time. Even though huge amounts of information may be stored in an enterprise, most of the time the decision-making rests on a comparatively smaller amount of data. One of the most important activities that a decision maker performs is to filter the existing information and find important and relevant data that can help in making a decision. Because the decision-making may require tapping the tacit knowledge of fellow employees, it is imperative for companies to develop systems and mechanisms to be able to do this (Payne, Bettman, & Johnson, 1993).

Existing research points to the fact that decision-makers act as satisfiers seeking a satisfactory solution rather than an optimal one. The rationality of individuals is limited by their cognitive limitations and the time available to make the decision (Simon, 1999).

1.2 Motivation

People need access to enough information at appropriate times for decision-making. Usually, the best alternative is to reach the right person who can provide the information that is needed. Knowledge management must become more social. The motivation of the present research stems from this primary reason. The modern knowledge worker requires the right source of knowledge for a given context, and at times this source may be an individual.

Another aspect of this is the knowledge that is never written down. It is known as tacit knowledge, which exists in individuals. Individuals who are interacting with other individuals understand this tacit knowledge, but for others, it is almost impossible to find (Davenport & Prusak, 2000).

Figure 2 depicts the evolution of organization systems to meet the needs of the evolving organization and group needs. A few years ago, systems were primarily read-only, in the form of web pages. Then, they became participative meaning that users contributed to the systems leading to a significant increase in the deployment of knowledge management systems. These systems were good enough to capture explicit knowledge in the form of documents created through collaborations but were completely insufficient for tracking tacit knowledge. As organizations became more aware, visual, and intelligent entities – because of their collective knowledge – organization systems also evolved to meet needs.

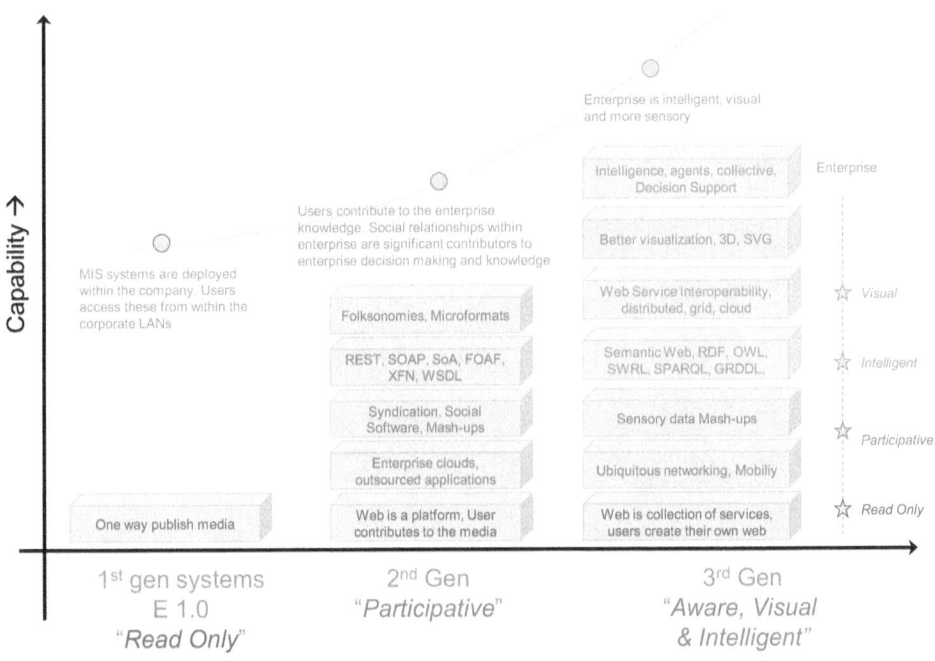

Figure 2 Organization system evolution through the years

The modern era is the era of knowledge. Enterprises, as well as other independent groups, are constantly striving to create and collate knowledge. Even though there has been an effort to build knowledge management systems, there are challenges because of the creators' disinclination to share knowledge, which is predominantly a social process, and must be supported by an organization's emphasis on knowledge sharing as a good group behavior (Connelly & Kelloway, 2003).

Knowledge workers depend significantly on generation and consumption of knowledge for their day-to-day operations. There is a significant effort in trying to codify knowledge to make it easier to retrieve when required. Apart from their staff, enterprises also depend on CoPs to build on the knowledge that they have. The information retrieval systems are mostly based on keyword-match techniques and suffer from information mismatching and overloading (Tao, Li, Zhong, & Nayak, 2008).

Organizations create and utilize knowledge in multiple ways. Figure 3 shows how organizations deal with knowledge. Of particular interest is the expert model and collaboration model, in which tacit knowledge plays an important part.

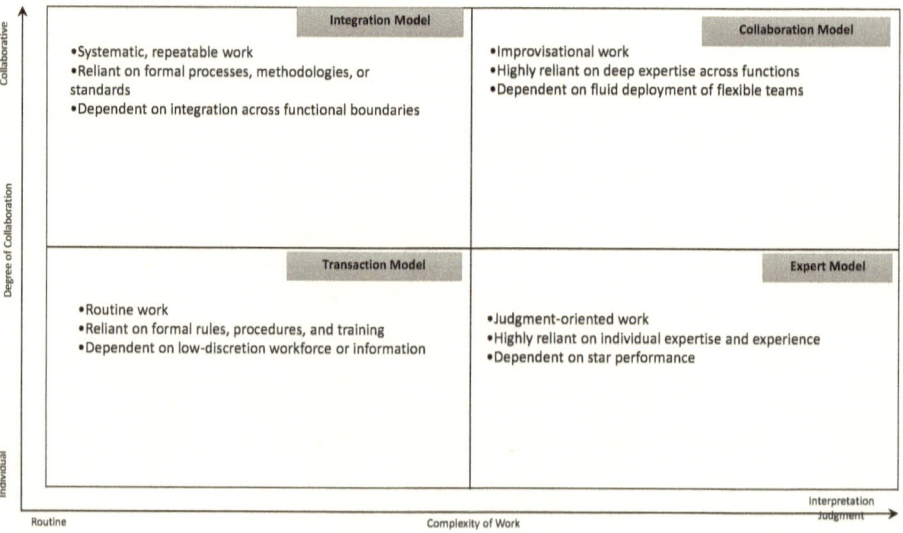

Figure 3 Relationship between collaboration and type of work (Davenport, 2005)

Organizations have found it difficult to gain access to all the knowledge that resides within. In the words of Lew Platt, CEO of HP, "Knowledge management is all that there is in our company. We live and die on our intellectual property, acquiring knowledge quickly, moving it around the company very quickly, it's all about knowledge transfer; starting with the customer" (Sieloff, 1999). The growing importance of knowledge in general and tacit knowledge in particular within businesses and CoPs presents a challenge to define methods that can improve upon the discovery of relevant sources so that it is possible to tap into tacit knowledge within organizations and CoPs.

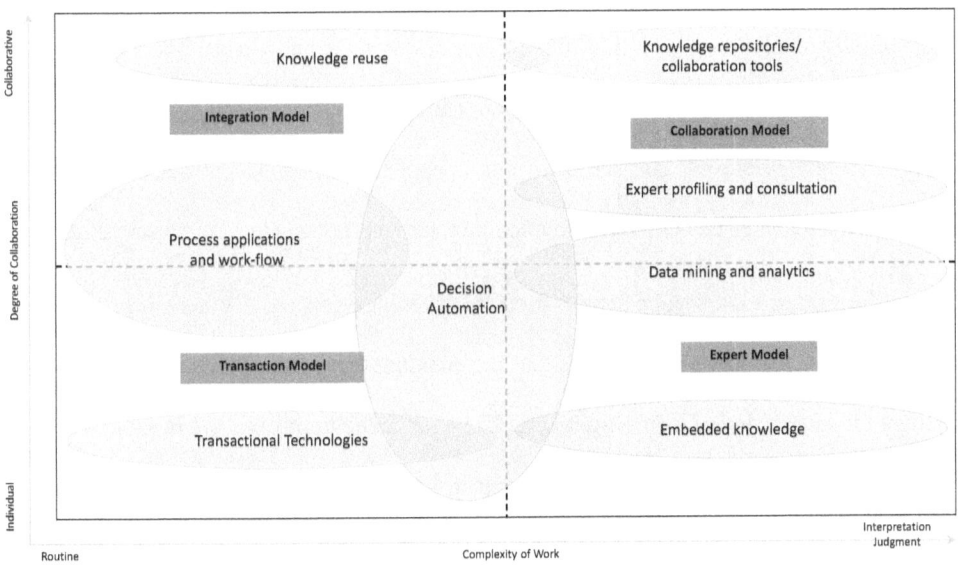

Figure 4 How individuals use knowledge (Davenport, 2005)

Figure 4 describes how organizations and CoPs utilize knowledge for day-to-day functioning. However, managing it becomes extremely important, and organizations must have a higher degree of collaboration, and the decision-making requires interpretation and judgment. Figure 4 illustrates activities in organizations based on the complexity of work and collaboration, and the top and bottom right quadrant is where the need for knowledge is felt most. The bottom right quadrant is also where tacit knowledge becomes extremely important.

1.3   Research Gap

To study knowledge management in general and tacit knowledge specifically, we need to identify knowledge workers and what knowledge they possess. We define knowledge worker as an individual who thinks for a living, and whose primary job is knowledge creation (Davenport, 2005).

This traditional definition is very restrictive. In the current environment, many need accesses to knowledge to perform their job better. Those who create or require access to knowledge to perform their job effectively are knowledge workers. The modern knowledge workers are complex individuals; they have multiple dimensions to their knowledge, and each of these dimensions is important in different contexts. They have various skills and interests, they control multiple resources, interact with many people in different contexts, and performs multiple tasks. Each of these is useful in tracking their tacit knowledge(Davenport, 2005).

High performers in a knowledge organization are different compared to other workers in many ways. One of the significant behaviors that differentiate them from others is their ability to leverage the knowledge of their co-workers. The knowledge in this context is explicit as well as tacit. Inter-personal skills that are inherent in individuals are important to leverage any knowledge; it becomes the only mechanism when we talk about tacit knowledge (Davenport, 2005).

As previously mentioned, knowledge management systems in organizations have limited success due to one or more of the following reasons:

- Most systems ignore the social dimension of such systems. It is important to capture relationships that are created between individuals due to their interactions.

- Keyword search systems form the foundation of current knowledge retrieval systems. This approach results in a mismatch of information.

- Most of the knowledge management systems have no way of identifying underlying communities.

- We did not come across any effort to track the evolution of knowledge over time. Knowledge of communities, groups, and individuals evolve over a period while the knowledge seeker may be interested in something that is not the top interest of the community. In conventional systems, such topics are buried under and impossible to find out.

1.4   Research Questions

The knowledge within individuals who form an organization or community impacts the knowledge contained within communities and an organization. The knowledge of an organization is a superset of organizational memory (J. P. Walsh & Ungson, 1991) and problem-solving skills of communities and individuals. Even though most organizational memory can be codified, there are still fragments of this knowledge that are stored as tacit knowledge in the minds of individuals. Similarly, many of the problem-solving skills are inherent to individuals in the form of tacit knowledge (Grant, 1996b).

The research problem concerns the discovery of explicit and tacit knowledge in the context of an organization and groups. Given the community structures, communication and knowledge assets within a community, an attempt is made to create a method and system that can discover knowledge.

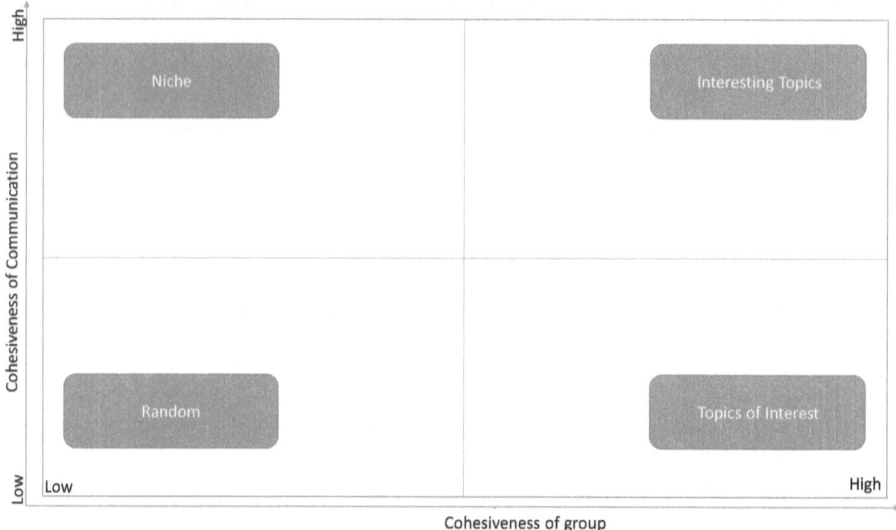

Figure 5 Classifying knowledge based on cohesiveness of groups and communication

Figure 5 classifies conversations among individuals based on the cohesiveness of group and communication. Topics of interest are particularly relevant in the context of knowledge sharing within communities and organizations. The following are the set of research questions examined.

1.4.1 Evolution of knowledge in a community and group

The first question that needs to be examined is how the knowledge about a topic evolves over a period. The exchange of information is considered within a group or community. What may be important to a group or a community at a point in time may not be important later. It is necessary to keep track of the evolution of knowledge because what is out of favor today may be required tomorrow, and as a group or organization, it should not be forgotten.

1.4.2 Evolution of knowledge of an individual

The second question that is examined concerns how the knowledge held by an individual evolves over a period. Importantly, it is seen very often in organizations that there is a sudden need for skillsets that are no longer current in the company. An ability to track the evolution of

20

knowledge would help find individuals who have dealt with the issues at least sometime in the past.

1.4.3 Formation of cliques

The third question that we examine concerns how the cliques form in communities. The flow of communication across individuals can identify these, and they can be utilized to draw useful inferences as to the expertise of cliques and communities.

1.4.4 Explicit and tacit knowledge

The fourth question addresses how a knowledge network can be used to locate the appropriate body of knowledge and its creator for a specific topic. The starting point of this discovery may be adjacent explicit knowledge which can be used to estimate tacit knowledge.

1.4.5 Representative phrases

The fifth question that is examined concerns finding the phrases of interest from a volume of communications and interactions.

1.5 Business and Managerial Implications

Organizations have suboptimal results consistently in their knowledge management endeavors because they treat the latter as a keyword search exercise. The failure of knowledge management exercises within organizations stems from the fact that knowledge sharing is an unnatural exercise, and in the absence of appropriate incentive systems, it is difficult to expect knowledge-sharing endeavors to be successful (Davenport, 2005).

The study proposes a new way to look at managing knowledge. We present a methodology that does not require any change in the way people work to extract knowledge from the existing communication repository. Systems are also deployed that can enable true asynchronous enterprise where the infrastructure makes relevant information available when a decision needs to be made. Emphasis is placed on the individuals. Knowledge sharing and dissemination are primary social activities, and the emphasis is on the social dimension. When an individual is evaluating options for a decision to be made, he or she can quickly know who are the best individuals in the organization or knowledge network to help.

Businesses and managers can also explore natural cliques that are created within the organization while in pursuit of knowledge creation. These natural cliques may point to islands of expertise in the areas of their work. It is best to leave them alone, but their expertise may be needed for decision-making in due course.

1.6 Contributions of the Work

This thesis contributes to the existing body of knowledge by defining a methodology to extract knowledge and sources of knowledge from an organization's communication and interaction repository. Organizations lose some of their knowledge when individuals leave an organization. The thesis describes the methodology and algorithm to build knowledge networks from a communication repository. A methodology is then defined to examine the evolution of individuals, groups, and communities over periods of time using the knowledge networks that were built.

This thesis contributes a methodology for using weighted knowledge networks within enterprises and CoPs to find the knowledge and to point to the individuals who may be the best candidates for tacit knowledge relevant to a given context. Knowledge adjacency is also proposed as a mechanism for finding phrases similar to those that somebody may be interested in.

The building of weighted knowledge networks allows the following inferences to be drawn:

-       Point to the most suitable individual for a community to talk about a phrase.

- Point to the most responsive individual in a community.

- Identify the best mediating individual in a community.

- Find the period when a phrase was of most interest.

When searching for a phrase in a body of knowledge, it may not be the best phrase for a given period. A methodology is proposed by which the best phrase and best period can be chosen to find the best expert or best mediating individual. Furthermore, the thesis explores the phenomenon of naturally forming cliques within communities and defines a methodology and algorithm to compute it. We also consider the impact of attrition of leading members of the community.

1.7 Definition of Knowledge

There is a clear distinction between knowledge and information. First, knowledge, unlike information, is about beliefs and commitment. It is a function of a stance, perspective, or intention. Second, knowledge, unlike information, is about action; it always has some end. Third, knowledge, like information, is about meaning. It is context-specific and relational. Nonaka also modifies the commonly accepted definition of knowledge, justified true belief, to argue that the truthfulness is not all that important. In his view, knowledge is nothing but "justified belief" (Ikujiro Nonaka & Takeuchi, 1995).

The 4I framework or organizational learning defines learning at three levels of individual, group, and organization and consists of four related sub-processes. It describes that individuals, groups, and organizations employ very different sub-processes to create knowledge and impart learning. An individual is mostly intuiting and interpreting while the group is integrating. The organization is institutionalizing. When we consider the groups, it is evident that the task of integrating is helped by the communication between individuals (Crossan, Lane, & White, 1999).

1.7.1 Ontology

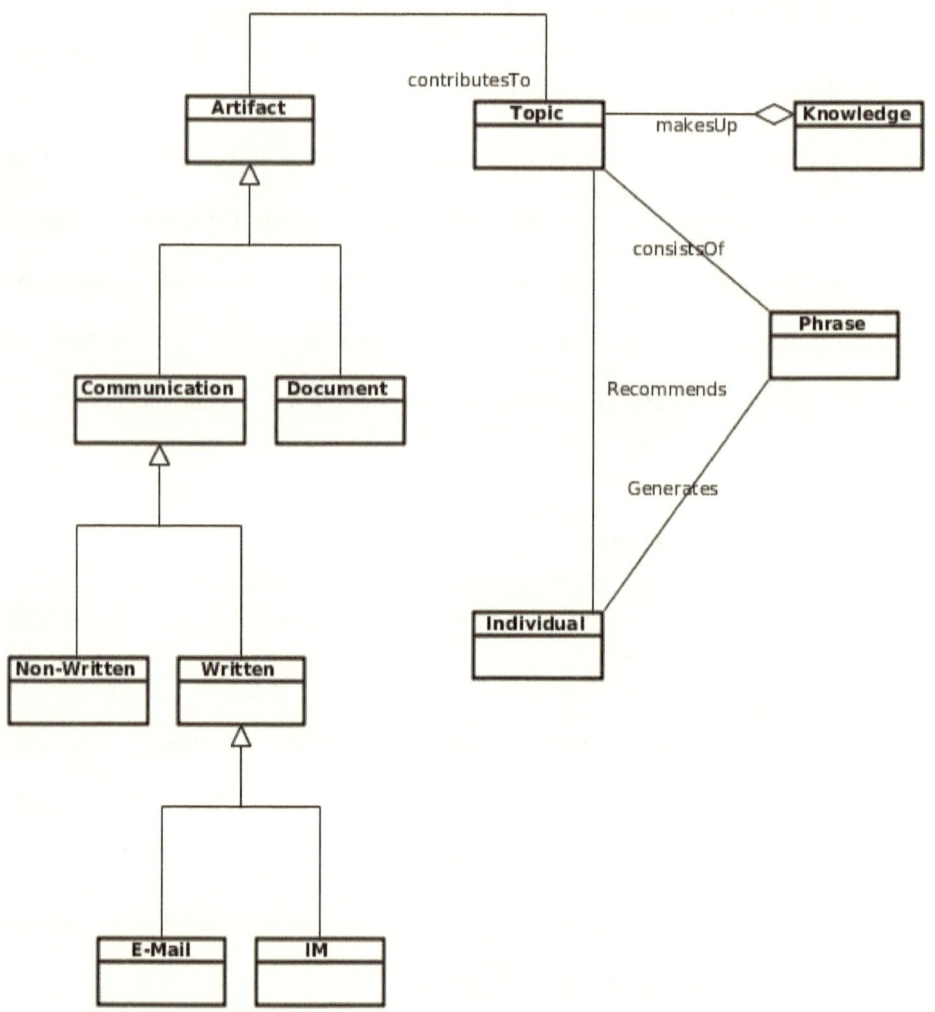

Figure 6 Knowledge ontology; knowledge is composed of topics contributed by individuals

In a Turing Lecture in 1976, Newell and Simon stated that there is no "intelligence principle," just as there is no "vital principle" that conveys by its very nature the essence of life. However, the lack of a simple *deus ex machina* does not imply that there are no structural requirements for intelligence. One such requirement is the ability to store and manipulate symbols. The physical symbol system has the necessary and sufficient means for general intelligent action.

The solutions to the problem are represented as symbol structures. A physical symbol system exercises its intelligence in problem-solving by searching, that is by generating and progressively modifying symbol structures until it produces a solution structure. The goal of this study is to build a system that can search for required knowledge and its source, the individual; it is of utmost importance to have a definition of knowledge that is relevant for this purpose. Organization communication repositories are the primary source of knowledge indicators; hence, knowledge is defined as represented by such repositories (Newell & Simon, 1976).

Partly inspired by Newell and Simon's Turing Lecture (1976), if intelligence is all about defining a symbol structure and heuristics search to find a solution structure, then knowledge in communication repositories could be well represented by tokens generated from this repository and a structure that represents the interaction between these symbols. To have a clear understanding of what is meant by knowledge, its ontology must be defined.

The model of knowledge is designed with the assumption that most of the codified knowledge is available in the form of unstructured data. Figure 6 describes the details of knowledge ontology for representation of knowledge and its components. In this model, the artifacts created by a community, in the form of communication, documents, etc., contribute to a topic. A topic is a top-level concept within which knowledge may be classified. A community would have one or at most few topics of interest. Few representative phrases identify these topics. Different individuals enrich topics by contributing phrases. The social dimension that is required for getting to tacit knowledge is captured by individuals recommending topic that may be another individuals' expertise.

1.7.2 Types of knowledge

Knowledge, at the top level, can be classified into two categories, namely, tacit knowledge and explicit knowledge.

1.7.2.1 Tacit Knowledge

Tacit knowledge is most difficult to track since it lies only in the minds of people who possess it. We can convert Tacit knowledge into explicit knowledge and vice versa(S. D. N. Cook & Brown, 1999). Since tacit knowledge is in the minds of individuals who possess it, the best way to get access to tacit knowledge is to be able to track and discover individuals who possess these. We have defined a methodology that can help us get access to the individuals possessing tacit knowledge.

1.7.2.2 Explicit Knowledge

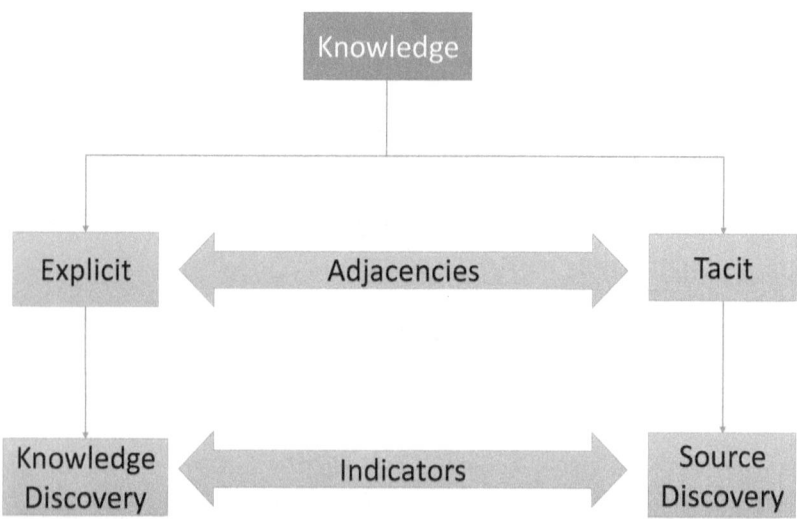

Figure 7 Knowledge types and their discovery; tacit knowledge is found in adjacencies of explicit by reaching to the source of knowledge

Vast volumes of explicit knowledge are available in the form of codified knowledge that is being created on a daily basis. The internet and company intranets are full of such knowledge.

In the case of explicit knowledge, the primary challenge is of knowledge discovery which manifests itself as an effort to discover knowledge in a form that is usable by an end user.

Even though we can't convert tacit knowledge to explicit and vice versa, their adjacencies are a very good indication of the presence of that knowledge. To accomplish this, we have defined a mechanism that helps build a relationship between tacit knowledge topics and explicit knowledge topics. These could be same or different at times.

Figure 7 describes the ways that we can reach knowledge. Knowledge discovery is the best mechanism to reach explicit knowledge. Keyword search-based systems are most appropriate for such situations. They can lead the user to the desired explicit knowledge that is stored in the organizations' IT systems. Source discovery is the most appropriate mechanism to get to tacit knowledge by looking at adjacencies. The best approach is the to find the person who has tacit knowledge and then through socialization; one can have access to that knowledge.

1.8   Organization of thesis

We have organized this thesis into multiple chapters as explained below.

- Chapter 1 contains an introduction to this thesis
- Chapter 2 provides a survey of existing literature related to the research; The existing body of knowledge relevant to the problem being examined is summarized.
- Chapter 3 discusses the work undertaken as part of this research. We define the flow of the information while it is being processed, the research methodology, dataset and, tools used.
- Chapter 4 presents an analysis of quantitative data obtained to validate usage of tacit knowledge within organizations. The hypotheses are tested related to important teams assigned to tacit knowledge in their day-to-day operations.

- Chapter 5 considers methodology that we have used to build knowledge networks from existing communication repositories. The evolution of knowledge networks and their members over periods of time is also examined.

- Chapter 6 defines the concept of knowledge adjacency and algorithm and methodology to compute it. There is further discussion of how natural cliques form within communities, which can be identified by the communication patterns across individuals.

- Chapter 7 concludes this study and includes the limitations and proposed future research.

- Appendix A describes the architecture of the system that was developed as part of this research work.

- Appendix B contains survey questionnaire that was administered as part of the exercise done in Chapter 4.

- Appendix C contains the process and detailed research related to literature survey performed.

- Appendix D has the table of results that we obtained from phrase representation analysis that we did as part group detection within the community.

- Appendix E has additional charts related to social network analysis of different topics.

- Appendix F contains SPSS tables from statistical analysis of exploratory research problem which is presented in Chapter 4.

- Appendix G enumerates the publication generated as part of this research.

CHAPTER 2

2. Literature Review

2.1 Knowledge, information, and data

The research questions that we have articulated are looking at communities and individuals. Knowledge, as defined by DIKW hierarchy, is closest to Tacit knowledge as defined by (Ikujiro Nonaka, 1991) and (Polanyi, 1966). We are looking at knowledge that is not codified and exists in the minds of individuals thus requiring us to locate the individuals with the knowledge. This is the reason why the focus of our research is knowledge rather than information or data.

The conventional wisdom is that data is a set of discrete objective facts about events. Data turns into information as soon as it is structured and given meaning. We interpret the information, put in context, or add meaning to it which yields knowledge. The hierarchy of data-information-knowledge should be turned around. Data emerge last—only after knowledge and information are available. There are no "isolated pieces of simple facts" unless someone has created them using his or her knowledge. Data can emerge if a meaning structure, or semantics, is first fixed and then used to represent information (Tuomi, 1999).

2.2 Knowledge representation

Representations constructed to emphasize particular, general, or interpretive knowledge can have differential impacts on managerial idea generation. Knowledge representations that can represent concrete and unambiguous knowledge are more effective in producing decision responses compared to general knowledge representations. Interpretive representations provide a means of combining abstract and concrete knowledge. Knowledge communication can be made more potent by figurative, ambiguous, and concrete representations(Boland et al., 2001).

Organizations often compete with each other under conditions in which relative position matters. The mixed contribution of knowledge to competitive advantage in cases involving competition for primacy creates difficulties for defining and arranging an appropriate balance between exploration and exploitation in an organizational setting(Nasierowski & Mikula, 2011).

Theory of conceptual dependency implies that in any language if two sentences have identical meaning, they should have only one representation and any implicit information in a sentence must be made explicit in the meaning(Shank & Abelson, 1977).

2.3 Organization learning

Learning in organizations can be defined as a process of detecting and correcting errors. The error is any feature of knowledge that inhibits learning. At primary level, individuals in each group view much of the information received from others as being inconsistent, vague and ambiguous. They also Do not want to discuss the quality of inconsistency, vagueness, and ambiguity. At the secondary level, we see secondary inhibitions which are the games that people play in order not to upset each other. Both the above factors reinforce each other, and eventually, they form a tight system that inhibits individual and organizational learning(Argyris, 1977).

Data are defined as symbols that represent properties of objects, events, and their environment.

They are the products of observation. But are of no use until they are in a usable (i.e., relevant) form. The difference between data and information is functional, not structural. Information is contained in descriptions, answers to questions that begin with such words as who, what, when and how many. Information systems generate, store, retrieve and process data. Information is inferred from data. Knowledge is know-how and is what makes possible the transformation of

- **Monopolies** –It results when only one person or one group holds knowledge that others need. The results are similar to other monopolies; the knowledge comes at a high price.

- **Artificial Scarcity** –A corporate culture in which knowledge hoarding is the norm, create scarcity.

- **Hoarding** –It is the barrier created by the possessiveness of the individual possessing knowledge. It could also result because of an executive who forbids sharing of knowledge.

- **The Not Invented Here** syndrome is the refusal to accept from others in the organization.

Knowledge generation within an organization consists of knowledge developed within the organization or acquired from outside. Many of the acquisitions in the industry are made with the specific purpose of acquiring knowledge. There is an organic connection of knowledge to individuals and the environment(Davenport & Prusak, 1998).

Figure 8 describes codification dimensions of different types of knowledge. Whether an organization or CoP needs to do something about a type of knowledge depends on its importance, what it needs to do depends on its type. Tacit knowledge is almost impossible to reproduce in a document. That is why codification process for the richest tacit knowledge in organizations is limited to locating someone with the knowledge, pointing the seeker to it and encouraging them to interact(Davenport & Prusak, 1998).

Tacit and ambiguous knowledge is especially hard to transfer from the resource that creates it to other parts of organizations. This type of knowledge is best transferred at water cooler conversations. Many Japanese firms have set up talk rooms to facilitate this kind of knowledge sharing. Encouraging after work group dinners and any other form of socialization among individuals within a company or group also facilitates such knowledge transfer. Explicit knowledge is easier to transfer. It is codified and can be easily made available through company

intranet and portals (Davenport & Prusak, 1998).

Figure 8 Codification dimension of knowledge (Davenport & Prusak, 1998)

Managing knowledge is difficult, here we are not only talking about typical knowledge

management systems but also the interpersonal and social side of knowledge transfer and

assimilation.

There are six ways of attaining knowledge(Montague, 1925). These are: -

- **Authoritarianism** – what we hold to be true

- **Mysticism** – intuition

- **Rationalism** – proving proposition by appealing to abstract and universal principles

- **Empiricism** –proving propositions by appealing to concrete and particular occurrences

- **Pragmatism** –finding meaning of an idea in its concrete results

- **Skepticism** –deny that any method can give genuine knowledge

- **Methods** – a combination of methods described above so that they supplement each

  other's defects.

There are three ways of interpreting knowledge(Montague, 1925). These are: -

- **Objectivism** –primitive, common sense, new objectivism. The essential contention of objectivism is that all objects of possible experience are also objects of possible existence.

- **Dualism** – sense data present in our consciousness and external things that can be inferred from those data. The contention of subjectivism is that all objects of actual existence are also objects of possible experience.

- **Subjectivism** – epistemological idealism. Dualism contends that separate causes determine the system of experienced objects and existing objects, and they vary independently of each other. Each expresses the situation from a particular angle.

To the extent that organizations exhibit characteristics of information processing, they should incorporate some memory, although not necessarily resembling human memory. In general, an organization may exist independent of particular individuals, but it should be recognized that individuals acquire information in problem-solving and decision-making activities. Through the process of sharing, the organizational interpretation system in part transcends the individual level leading to an organization preserving the knowledge of the past even when key organizational members leave. In its most basic sense, organizational memory refers to stored information from an organization's history that can be brought to bear on present decisions. This information is stored as a consequence of implementing decisions to which they refer, by individual recollections, and through shared interpretations. Organizations retain information in individuals, culture, transformations, structures, ecology, and external archives. Only individuals have the ability understand the "why" of a decision in the context of organization's history. Organization's culture as an aggregation of individuals' shared beliefs, also reflects information about the *who*, *what*, *when*, *where*, and *how* of a decision stimulus and response. Transformations, structures, and ecology, however, might not retain information about a decision stimulus but they inhabit an organization's response to such a stimulus(J. Walsh & Ungson, 1991).

State of the art systems uses advanced data mining and artificial intelligence techniques to deploy systems that can handle organizational knowledge and generate intelligent inferences out of it (Ribino, Augello, Lo Re, & Gaglio, 2011).

Organizations must make interpretations. Managers literally must wade into the ocean of events that surround the organization and actively try to make sense of them. Organization participants physically act on these events, attending to some of them, ignoring most of them, and talking to other people to see what they are doing. Interpretation is the process of translating these events, of developing models for understanding, of bringing out the meaning, and of assembling conceptual schemes among key managers. This process can be seen as operating in three stages. The first stage is scanning which is defined as the process of monitoring the environment and providing environmental data. In the second stage, the human mind is engaged, perceptions are shared, and cognitive maps are constructed, an information coalition of sorts is formed, thus the organization experiences interpretation when a new construct is introduced into the collective cognitive map of the organization. The third stage is learning, which is defined as the process by which knowledge about action outcome relationships between the organization and the environment is developed(Daft & Weick, 1984).

Knowledge-based theory of the firm defines organizations consisting of multiple individuals exist because there are gains from specialization in knowledge acquisition and storage. Given the efficiency gains of specialization, the fundamental task of organization is to coordinate the efforts of many specialists. Although widely addressed, organization theory lacks a rigorous, integrated, well-developed and widely agreed theory of coordination. A knowledge-based view of the firm encourages us to perceive interdependence as an element of organizational design and the subject of managerial choice rather than exogenously driven by the prevailing

38

production technology. Specialized knowledge can be integrated using four mechanisms, Rules and directive, Sequencing, Routines, Group problem solving and decision making. Knowledge integration requires different types of knowledge. Language, Symbolic communication, Commonality of specialized knowledge, Shared meaning, Recognition of individual knowledge domains(Grant, 2013).

An organizational learning context consists of organizational culture, structure, and infrastructure. The culture existing within the learning organization places great emphasis on learning and knowledge, creating an atmosphere of trust within which individuals feel empowered to experiment with new approaches to business, often resulting in the development of new core competencies. The culture consists of vision, leaders, desire for continuous improvement, the value of knowledge, etc. The solution to many of these problems comes in the form of flatter organizational structures with reduced cross-functional boundaries. Because the development of new knowledge is dependent on the interchange of ideas between specialists in the same field, there is also the need to establish various functional groupings. Conversely, organizational knowledge must be holistic to ensure that specialist knowledge from related areas is fully integrated. Explicit knowledge is comparatively easy to store and communicate, with most media being effective. However, the abstract and intangible nature of implicit knowledge makes this a far more difficult proposition(Pemberton & Stonehouse, 2000).

In multi-unit firms, projects in divisions with short network path lengths to other divisions that possessed related knowledge obtained more knowledge from other divisions and were completed faster, likely because of the search benefits accruing to project teams with this network position. In contrast, neither the extent of available related knowledge in the Company nor the path length in the entire network explained the amount of knowledge obtained from other divisions and project completion time(M. Hansen, 2002).

Knowledge creation processes are related to organizational learning playing the key role in improving organizational performance(Ramírez, Morales, & Rojas, 2011).

2.6   Knowledge in groups and communities

Individuals possess various bits of knowledge in their respective fields, but the "body of knowledge" of any field is possessed by groups, not by individuals. Put another way, the body of knowledge of a group is "held in common" by the group. We do not expect every individual in a group (discipline, profession, craft, etc.) to possess everything that is in the "body of knowledge" of that group (in fact, this is likely to be impossible, unnecessary, and perhaps even undesirable). The body of knowledge is possessed by the group as a whole and is drawn on in its actions, just as knowledge possessed by an individual is drawn on in his or her actions. For our purposes, the term "practice" can be understood as referring to the coordinated activities of individuals and groups in doing their "real work" as it is informed by a particular organizational or group context. Knowledge is about possession while knowing is about one's interactions with the things of the social and physical world(S. D. N. Cook & Brown, 1999).

Communities of practice and their related technology support systems play an important role for individual-level knowledge in large organizations, as did the team-level attributes of transactive memory and absorptive capacity. Face to face opportunities can be used for transfer of tacit knowledge(Griffith & Sawyer, 2010a).

Unlike other organizational groups, communities of practice are organic. At present, we have little experience as to how to oversee and harness value from this type of organizational entity(Smith & McKeen, 2004). What we do know is that communities differ from other organizational groups, especially teams. Communities are most often confused with teams. But unlike teams, communities are typically voluntary and unstructured groups with a membership that cuts across internal and external organizational boundaries. While focus group members were inclined to suggest that teams and communities had many areas of overlap, researchers in this area believe that it is important to understand the principal differences between the two. The advent of communities signals that organizations are becoming more fluid than they have been in the past. Whereas organizations have traditionally had clear borders, in the future these will become more and more 'fuzzy' as communities reach out into their members' professional communities and include the enterprise's allies, partners, vendors, and customers. Internally, the organization will look more like webs of participation, again crossing traditional project and functional boundaries.

Communities form the foundation of knowledge management because it is through them that knowledge gets both created and turned into action. While they have been made possible by technologies that enable people to share insights and ideas around the world, they are first and foremost a social mechanism(Smith & McKeen, 2004).

Only 15-20 percent of members within a community of practice are core members; if any of them leave, there is a significant risk of loss of knowledge within that community(Lee, Suh, & Lee, 2014).

Knowledge networks require the convergence of three areas: communities of practice, knowledge assets, and technology(Neumann & Prusak, 2007).

2.6.1 Communities

2.6.1.1 What are communities

Important to knowledge discovery are the organizations, groups, and communities that lie within the organization. Groups and its tasks are sanctioned by the organization while the organizations do not recognize the communities. Groups are canonical, bounded entities that are organized or at least sanctioned by the organization and its view of tasks. The communities, on the other hand, are often non-canonical and not recognized by the organization. They are fluid and often cross the boundaries of the organization (Brown & Duguid, 1991).

Communities of practices are as a mutual engagement in a joint enterprise with a shared repertoire(Etienne Wenger, 1999).

Communities of practice are not self-contained, rather they develop in larger contexts with specific resources and constraints. These are groups of people informally bound together by shared expertise and passion common goal(E. C. Wenger & Snyder, 2000). Communities of practice are different from formal work groups, projects teams and informal networks in the sense that the primary purpose of communities is the building and exchange of knowledge. Delivery of a product of accomplishing a task is seldom the goal of communities of practice. The advent of communities is also a signal that organizations are becoming more fluid than they have been in the past(Smith & McKeen, 2004).

In virtual teams, knowledge sharing gets impacted by many factors. An effective reputation system needs to be a key aspect of any virtual community of practice. Since most of the members rarely meet each other, there needs to be an effective way of building a reputation system which can be used virtually(Emelo, 2012).

2.6.2   Cliques

The informal communication system is sometimes used by organizational members to advance their aims. From this arises the phenomenon of cliques – groups that build up informal network communications and use this as a means of security power in the organization. These cliques form because different subsets of individuals within the community may be interested in specific areas of a topic that the community represents. They become islands because there is little communication happening across the groups. Individuals that form these cliques may be part of multiple such cliques, but their persona and communication pattern within a clique is very different than when they participate in a different clique(Simon, 1950).

2.6.2.1 *Clique Formation Algorithm*

Clique detection within a knowledge graph is accomplished by finding clusters of nodes. Several metrics have been conceived to distinguish one network from another.

For example, betweenness of a link is the number of shortest paths between pairs of nodes that include the link. Modularity is another measure that is used to define the cohesiveness of a network. Networks with high modularity would have higher connectivity within the nodes in a module while lower connectivity in nodes across the modules. It is calculated as a fraction of the edges that fall within a given group minus the expected fractions in the case of randomly distributed edges.

One way of detecting underlying cliques within a graph is done by a divisive hierarchical algorithm, in which links are iteratively removed based on the value of their betweenness. In its most popular implementation, the procedure of link removal ends when the modularity of the resulting partition reaches a maximum(Girvan & Newman, 2002).

The problem of cluster structure detection can also be defined as the problem of compression of information. The stated goal of such compression is to be able to recover a structure that is as close as possible to original structure(Newman, Barabasi, & Watts, 2006).

Another way of detecting cliques is by partitioning the network into networks of densely connected nodes with nodes belonging to different cliques only sparsely connected(Blondel, Guillaume, Lambiotte, & Lefebvre, 2008). It is a multi-step technique based on (Girvan & Newman, 2002). After a partition is identified in this way, cliques are represented by new nodes which are also known as supernodes, yielding a smaller weighted network. The procedure is then iterated until modularity of the original graph does not increase any further. For our analysis, we have used (Blondel et al., 2008).

2.7 Knowledge in individuals

While looking at the managing the knowledge, it is important to find how people understand. Theory of conceptual dependency implies that in any language if two sentences have identical meaning, they should have only one representation and any implicit information in a sentence must be made explicit in the meaning. They also provide a framework or language representation. Meaning propositions underlying language are called conceptualization. Conceptualizations could be active or stative; once a conceptualization is active; it takes the form of Actor Action Object Direction. Conceptualization in the stative state has the form Object (is in) State (with value) (Shank & Abelson, 1977).

Memory sharing in close relationships can be explained by providing an analogy to two computer systems that need to share data without having shared memory. These computer systems essentially depend on a kind of directory that provides some lookup for memory that is owned by the other computer system. Each person in the relationship learns what other person knows and maintains a directory for each other's knowledge. The processes of the building of this directory start with social categorization and stereotyping, and this directory is updated as individuals in the relationships interact(Wegner, Erber, & Raymond, 1991).

Organization decision making can be aided by providing most relevant information to decision makers. Given the fact that organizations today are deluged with data, knowledge, and information, it is of utmost importance to present what is required for a particular decision. There are instances within organizations where the information needed for a particular purpose is available within the organization but the group (or the individual) that needs it doesn't know where it is or how to access it. The concept of transactive memory can be effectively applied to groups and organization to make sure that the organizational memory is available for the whole organization. The information available to an organization can be chunked into a variety of domains. Employees specialize in one or two domains and maintain directory about knowledgeable people in other domains. When needed, information from individuals specializing in different domains is retrieved through communicative transactions(Anand, Manz, Glick, & Glick, 1998).

The acquisition does not guarantee the knowledge to the buyer organization. More often than not, individuals because of which acquisition was made, end up leaving because of disruption in organization processes and networks. Companies also acquire knowledge by renting it; organizations should take steps to retain knowledge that was acquired by renting. Since most of the knowledge resides in the minds of people in the form of tacit knowledge, organizations make an effort to codify this knowledge to help in transfer and sharing of knowledge (Davenport & Prusak, 1998).

Movement of knowledge within the organization requires the movement of the specialists who possess it; then effective knowledge utilization will tend to require that individuals occupy multiple organizational roles involving membership of multiple teams. If the primary resource of the firm is knowledge, if knowledge is owned by employees, if the individuals who possess it can only exercise most of this knowledge, then the theoretical foundations of the shareholder value approach are challenged(Grant, 1996a).

The social context of organizational learning has two distinctive features. The first is the mutual learning of an organization and the individuals in it. Organizations store knowledge in their procedures, norms, rules, and forms. They accumulate such knowledge over time, learning from their members. At the same time, individuals in an organization are socialized to organizational beliefs. Such mutual learning has implications for understanding and managing the trade-off between exploration and exploitation in organizations. The second feature of organizational learning considered here is the context of competition for primacy(March, 1991).

2.8 Knowledge sharing and diffusion

Knowledge sharing and transfer are extremely important in the context of organizations, and it is hard across functional boundaries. Three approaches; syntactical, semantic and pragmatic can be used to minimize the problem of knowledge transfer across the functional boundary (Carlile, 2002).

- The syntactical approach is based on the existence of a shared and sufficient syntax at a given boundary. A sufficient syntax is efficient because differences and dependencies have been specified and agreed to in advance.

- The semantic approach recognizes that differences exist or emerge over time, so individuals have different interpretations of a word or an event. In this way, the semantic approach recognizes that there are always differences in kind and the emergence of novelty on one or both sides of the boundary is a natural outcome in settings where innovation is required. From this approach integrating devices should be seen as processes or methods (i.e., standardized forms and other shared methods) for translating and learning about the differences and dependencies at a boundary.

- The pragmatic approach recognizes that differences in knowledge are not always adequately specified as differences in degree or interpretation, but that knowledge is localized, embedded, and invested in practice. This pragmatic framing of knowledge highlights the negative consequences that can arise given the differences and dependencies at a boundary.

In many cases contributing to knowledge management system is perceived as a loss of personal expertise while accessing such a system is seen as a sign of inadequacy. Also, there is a lack of effective mechanisms to distill knowledge from debriefs and discussions(Chua & Lam, 2005).

When transferring knowledge across strategic partner, both knowledge-specific variables (i.e., tacitness and complexity) and partner-specific variables (i.e., prior experience, cultural distance, and organizational distance) impacted this process(Simonin, 1999).

Trust may be a condition to knowledge sharing but does not have a positive effect on the sharing of knowledge per se. Although the absence of trust may impede people's motivation to share knowledge with others, it is unlikely that those who have high levels of trust in others are more likely to share knowledge than those with moderate trust levels. Members of teams in which members have been together for a longer time tend to share more knowledge between one another than members of younger teams. Longer-lived cooperative and knowledge sharing networks tend to become increasingly strong at knowledge sharing over time. Knowledge sharing social capital thus appears to be couched in the membership of experienced teams, rather than in the level of trust between individual members(Bakker, Leenders, Gabbay, Kratzer, & Engelen, 2006).

Knowledge sharing is human behavior that is influenced by both the knowledge sharing environment and other knowledge workers in the environment. Knowledge workers are diverse and heterogeneous. The KM models and tools that exist today do not address the heterogeneous attributes of the knowledge workers and pay minimal attention to the interaction between knowledge workers(Small & Sage, 2006).

Planned social interaction is not conducive to sharing. Interpersonal relationships and mutual understanding among individuals are better enablers for sharing of knowledge(Yang, 2009).

Knowledge sharing can be (also) a self-interested behavior; that is to say, a behavior that leads professionals to help themselves while – and before – helping their organization. Higher propensity and capacity to promote and implement new ideas, in fact, was facilitated by active involvement in knowledge sharing behaviors. Furthermore, sharing best practices and sharing mistakes are two distinct behaviors that professionals should seek to accomplish. Organizations need to establish a culture where individuals feel comfortable in sharing the mistakes because these contribute to the process of continuous improvement(Mura, Lettieri, Radaelli, & Spiller, 2013).

Structured practices form the foundation of individual action and social relationships on the shop floor, creating a shared language and knowledge base that support the sharing of tacit knowledge. An engaging environment not only improves performance but also favors knowledge dissemination. An engaging environment is supported by shared language and knowledge, which are developed through intense communication and facilitated by both a strong sense of collegiality and a social climate that is dominated by openness and trust(Nakano, Muniz, & Batista, 2013).

Eagerness and willingness to share are positively related to knowledge sharing. An agreeable style is positively related to team members' willingness to share their knowledge, whereas an extrovert communication style of a team is positively related to both eagerness and willingness to share. Performance beliefs and job satisfaction are both related to the willingness and eagerness to share knowledge(de Vries, van den Hooff, & de Ridder, 2006).

Knowledge practices extend beyond organizational boundaries. Information needs and acquisition, and knowledge sharing, creation and reuse do not follow organizational boundaries. Operational knowledge sharing helped in getting the work done; strategic knowledge sharing helped in understanding future challenges and defining objectives when preparing strategies and recommendations. At the personal level, through knowledge sharing, participants boosted their professional development(Lahtinen, 2013).

2.9 Knowledge creation by contingent workforce

About professional and technical work, contingent workers may have stronger public skills (such as industry- or occupational- based skills) than individuals who remain in relatively few organizations throughout their careers(Matusik & Hill, 1998). Because contingent workers use their skills in numerous organizational settings, their depth of knowledge about public skills is increased. Also, since the variety of one's experiences is one of two factors influencing the quality of an individual's knowledge (Ikujiro Nonaka, 1994), contingent workers are likely to have higher knowledge levels in industry and occupational best practices compared to their counterparts who remain in one or a few organizational settings. Contingent work may disseminate valuable private knowledge into the external environment, leading to the decay of competencies and a loss of competitive advantage. The departure of individuals causes a loss to the firm routines and practices in which these individuals participated. Contingent work also may stimulate the creation of new component knowledge. Exploration, as well as exploitation activities, are important in the quest to continually build knowledge. Exploration involves trying new processes and developing ideas that are outside of an organization's repertoire of routines(March, 1991).

Knowledge sharing is directly, and positively, connected with sharers' higher propensity to promote and higher capacity to implement new ideas. In doing so, the results allow highlighting and better clarify the direct benefits that the sharer of knowledge could attain through knowledge sharing – i.e., higher innovative behavior. Sharing best practices can trigger both the promotion and implementation of new ideas(Mura et al., 2013).

2.10 Tacit and explicit knowledge

Knowledge is an epistemology of possession while knowing is an epistemology of practice. Knowledge can be explicit and tacit and individual or group. Tacit knowledge cannot be turned into explicit, nor can explicit knowledge be turned into tacit. Tacit knowledge is acquired on its own; it is not made out of explicit knowledge. Before being generated, one form of knowledge does not lie hidden in the other. Knowing is to interact with and honor the world using knowledge as a tool. Engaging in conversation about knowledge with another individual is a practice that does epistemic work; it is a form of knowing. Knowing entails the use of knowledge as a tool in the interaction with the world. This interaction, in turn, is a bridging, a linking, of knowledge and knowing. And bridging epistemologies makes possible the generative dance, which is the source of innovation. The production of new knowledge does not lie in "a continuous interaction between tacit and explicit knowledge" but rather in an individual's interaction with the world. Specifically, it lies in the use of knowledge (explicit and tacit) as tools of productive inquiry (of the sort we have called "knowing") as part of our dynamic interaction with the things of the social and physical world(S. D. N. Cook & Brown, 1999).

If all knowledge is explicit, then we can't look for a problem or look for a solution (Plato & Jowett, 2012).

We can know more than we can tell. Tacit knowing always involves two kinds of things. In the act of tacit knowing, we attend from something for attending to something else; the first term is this relationship is proximal and the second term is distal. Polanyi defines that tacit knowing has four aspects(Polanyi, 1966). These are –

- the **functional** aspect aspects of the knowledge consist of a set of elementary movements for their joint purpose

- the **phenomenal** aspect can be defined by the fact that we are aware of the proximal term of an act of tacit knowing in the appearance of its distal terms

- the **semantic** aspect can be understood by the fact that we attend to the meaning of the impact of our actions on our hands in terms its effect on the things to which we are applying it

- the **ontological** aspect of tacit knowing explains what tacit knowing the knowledge of.

The efforts to formalize all knowledge to the exclusion of any tacit knowing are self-defeating, therefore, if problems exist, and discoveries can be made by solving them, we can know things that we cannot tell(Polanyi, 1967). Tacit knowing is shown to account

- for a valid knowledge of a problem.

- For the scientist's capacity to pursue it, guided by his sense of approach.

- For valid anticipation of the yet indeterminate implications of the discovery arrived in the end.

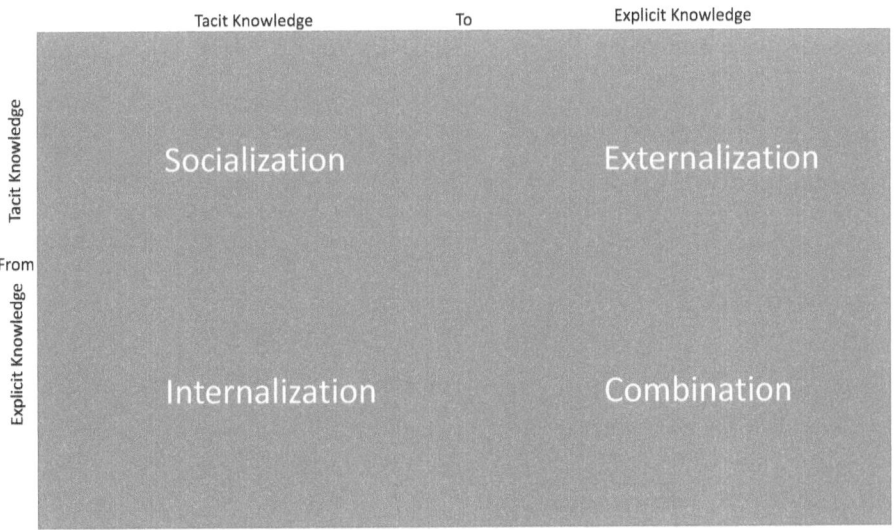

Figure 9 Modes of knowledge creation and transfer for tacit to tacit or explicit and vice versa (Ikujiro Nonaka, 1991)

Figure 9 describes four ways knowledge can be created and transferred. Tacit knowledge can be created and transferred among individuals through shared experiences arising out of social interaction, at times it is not even necessary to use a common language. Explicit knowledge can be created and transferred by sharing and transferring existing body of knowledge. Since explicit knowledge is kept in a physical form (documents etc.), it can be transferred, modified, combined and shared across individuals and groups. The other two types of knowledge creation and transfer involved the transformation of knowledge from tacit to explicit or vice-versa implying that both the tacit and explicit knowledge are complementary and can be converted to one another. An individual in an organization may internalize existing explicit knowledge, and it becomes tacit knowledge for him while at the same time, in some other situations, an organization may decide to codify existing tacit knowledge within the organization and make it explicit. This process is called externalization(Ikujirō Nonaka, 1994).

The use of tacit knowledge is assessed with special emphasis on its significance and implications in the innovation process. The following problem areas can, therefore, be categorized into individual and collective factors(Ikujirō Nonaka, 1994),

- the organization structures
- the culture of companies
- unclear goals
- unclear incentives
- wrong authorities
- the physical layout of companies.

Individual factors are (Ikujirō Nonaka, 1994)

- The readiness of every individual to share, e.g., if knowledge is a source of power and prestige.

-   The social culture of the company.

Collective factors relate to the fact that individual experiences should be transferred and acknowledged to a broader base within the firm. The culture of the enterprise strongly influences the transfer activities. A precondition to making this happen is to ensure a climate of openness and trust. During the transfer of tacit knowledge, communication problems at the interface can arise because many companies focus on the specialization of work. An increase in the degree of specialization results in higher isolation and narrower the perspective within a firm(Seidler-de Alwis & Hartmann, 2008).

Tacit knowledge is used to perform the skills since it remains residents in mind. Codified knowledge is expressed as information using systems of symbols; encapsulated knowledge is knowledge embedded in the physical assets such as machines or products(van den Berg, 2013).

While the transfer of explicit knowledge is associated with the firms' willingness to take the risk, transfer of tacit knowledge is intimately related to high trustworthiness(Becerra, Lunnan, & Huemer, 2008).

Chapter 3 defines the boundary of work that we propose to do to answer research question enumerated in 1.4. We also provide details of the information model, the research methodology, dataset and computing infrastructure that was used for this purpose.

CHAPTER 3

3. Proposed Work

The process begins with a dataset that is similar to the communication within a CoP which is then converted to a knowledge network so that social network analysis algorithms can be applied to it. Once these knowledge networks are available, they are analyzed further to determine their evolution.

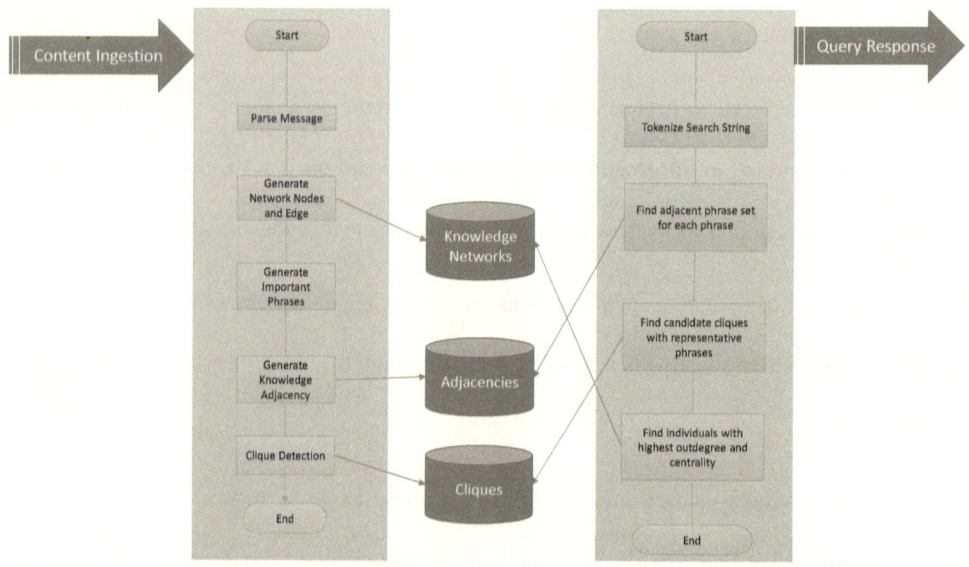

Figure 10 Knowledge capture and retrieval workflow from content ingestion to appropriate query response

Figure 10 describes the overall workflow involved in the processing of communication archives and a query interface to extract the source of the knowledge. The workflow consists of two broad sub-workflows. Messages are received and ingested in the system, and when this takes place, a sequence of processes operates on it, which results in a knowledge base. This knowledge base is later used by the query interface of the system to provide appropriate responses.

3.1 Scope of work

3.1.1 Research boundary

For this research, only textual interactions among users would be consumed. Other organization sources like multimedia content, web-based systems will not be consumed and processed. The defined architecture will provide an architecture that can handle all these sources, but these are not implemented because it is extremely difficult to get data from such sources. These could be taken up as future enhancements.

3.1.2 Message ingestions and Network Building

The messages are ingested into the system and processed. The output from this phase of work is a knowledge network. The construction of knowledge network is defined in Chapter 5.

3.1.2.1 Message parsing

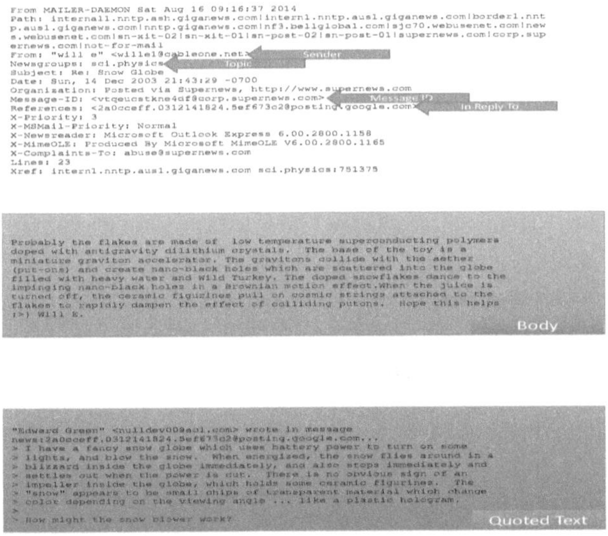

Figure 11 Format of a Usenet message; consists of metadata in form of headers, body and quoted text

Figure 11 describes the format of a Usenet message. As can be seen from the message structure, the metadata points to the sender, receiver, and additional information that helps in deciding which messages respond to each other.

When the system receives a message, it is processed in a series of steps to generate processed data that can be useful for a query processing system.

The first step is to process each message that is part of the dataset and extract the relevant information. It is necessary to understand the structure of a message to convert this unstructured message into something that contributed to the formation of the knowledge network.

3.1.2.2 *Generating knowledge network*

We process each of the messages from the repository and then construct and knowledge network with following steps.

- Each person is considered a vertex
- Each message that is a reply to an incoming message is considered an edge towards the original from the individual that replied
- Each edge is weighted by the importance of phrase
- There could be multiple directed edges between two individuals

3.1.2.3 *Generate important phrases*

The next step is to identify important phrases that are part of the body of these messages. The important phrases, after the process of de-duplication, become the candidates for edge labels.

3.1.3 **Generate adjacencies**

The identified phrases are then run through a process of adjacency identification. These adjacencies are important inputs to finding appropriate keywords while querying the knowledge networks. This is described in detail in Chapter 6.

3.1.4 Clique detection

The last process of ingestion workflow identifies inherent cliques within the knowledge networks. It also identifies the candidate phrases related to the cliques. This is also described in detail in Chapter 6.

3.1.5 Knowledgebase

The system generates an interim database that is used for query processing. This knowledge base consists of the following three types of information.

- Knowledge networks – This consists of definitions of knowledge networks, edges, and nodes that are derived from the communication archive.

- Adjacencies – This consists of adjacency definition for each important phrase identified in the system

- Cliques – This consists of the definition of cliques identified in the knowledge networks

3.1.6 Query processing

End users belonging to a CoP or an organization can use the query processing interface to find the knowledge and its source that they interested in.

3.1.6.1 Tokenize search string

An end user interested in querying the system enters a search string, this search string is processed, and only relevant phrases are kept for further querying.

3.1.6.2 Adjacent phrases

Before initiating a search with a keyword, the system also finds adjacent phrases. This augmented list of keywords is used to query the knowledge base that was generated as part of ingesting.

3.1.6.3 Finding cliques

Based on the augmented keyword list, the system finds the cliques that may be the best candidate to find the source of knowledge.

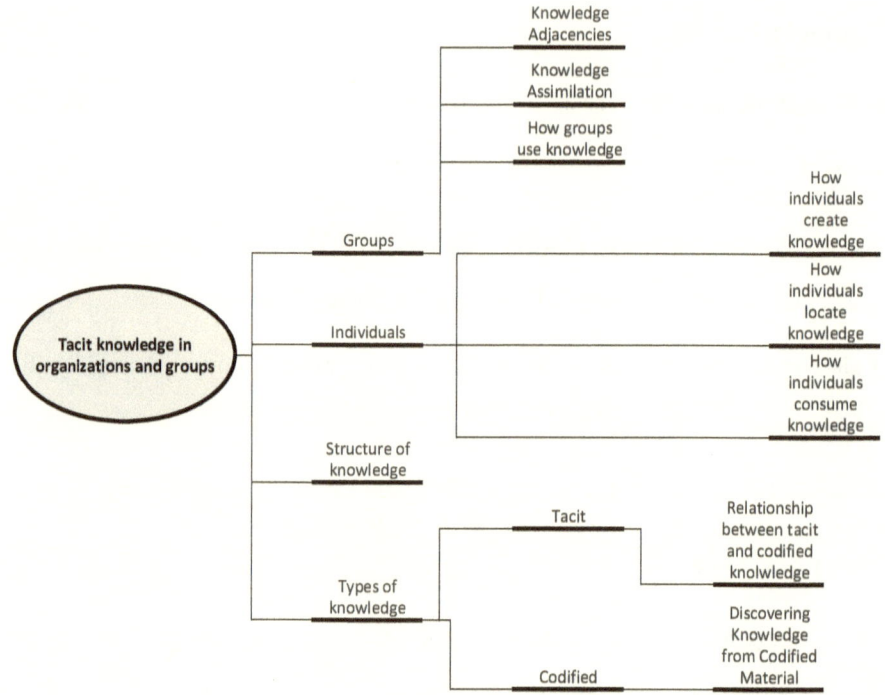

Figure 12 Knowledge in groups and communities, important questions that need to be asked

3.1.6.4 Finding the best candidate

The list of candidate cliques is used to find the individuals with the highest outdegree, indegree, and betweenness. These are individuals who are of interest. Figure 12 enumerates the different facets of knowledge. It is considered how individuals create and consume knowledge, how groups assimilate and integrate it. It is also relevant to determine how tacit knowledge may be related to explicit knowledge. Knowledge discovery is one aspect of the knowledge that is extremely important

3.2 Information model

The tacit knowledge of an individual resides in his mind; it is a distillation of experiences within the organization where he or she is working. Social structure is a good proxy to understand the tacit knowledge that an individual possesses. Communication with others, colleagues' perception of the individual, and the work that an individual is doing are elements of tacit knowledge. In this section, a model is proposed that defines the relationship between the interaction by an individual with others and the contribution of such interaction to knowledge networks.

3.2.1 Information exchange

Information exchange is the information content that the person has exchanged with other people. The information exchange always has a valid set of finite receivers of information.

$$I_t = \sum_{m=1}^{M} i_{t,m}$$

Where $i_{t,m}$ is the information content derived out of message m about topic t. For example $i_{t,m}$ can be calculated by Latent Semantic Analysis(Alias-i, 2008) where rows of input matrix represent topics and columns represent each instance of information exchange.

3.2.2 Recommendation

The recommendation is a specific type of information exchange when another person B identifies a person as an expert on the relevant topic. The information value of a recommendation is proportional to the recommenders' expertise in the topic, e.g., if the recommender has high expertise level and is still recommending another person in that topic, the information has more value than somebody who has low expertise and is recommending another person. Every instance of a received recommendation would have a positive outcome as regards the expertise of the receiver of recommendation.

$$R_t = \sum r_{t,i}$$

Where $r_{t,i}$ is the information content of ith recommendation received by the person about topic t. This is calculated in a fashion very similar to $i_{t,m}$ except that only content that consists of recommendations is taken into account, e.g. e-mail forwards

3.2.3 Broadcast

Broadcast content is generated by the user and published. There is no specific receiver.

$$B_t = \sum b_{t,i}$$

Where $b_{t,i}$ is the information content of ith instance of the content published by the individual about topic t. This will also be used using LSA.

3.2.4 Activities

Activities performed in the past by a specific user indicate that individual has expertise in topics related to that activity. There is no specific receiver.

$$A_t = \sum a_{t,i}$$

Where $a_{t,i}$ is the information content of ith instance of the activities performed by person related to topic t. This can also be calculated using LSA.

3.2.5 People expertise

The expertise of people in a group/community is made up of the following components.

1. Information exchange between people
2. Recommendation exchange between people

60

3. Information broadcast by individuals to the world

4. Activities performed in the past by people

$$E_t = f(I_t, R_t, B_t, A_t)$$

3.3 Research methodology

This research was carried out in multiple steps. The methodology combined quantitative research, social network analysis, and text mining of the dataset. A quantitative study was undertaken to validate hypotheses related to the impact of individuals leaving a community and group. The UTZOO Usenet Archive (Spencer & Wiseman, n.d.) was used to analyze the communication patterns across individuals and to build knowledge networks and graphs. Once the graphs were generated, they were analyzed using network theory and inferences are drawn.

Text mining techniques were used to extract interesting phrases; the similarity of texts enriched the retrieval mechanisms. Statistical and machine learning techniques were used to identify and build patterns in data over time. Additionally, interesting phrase detection was conducted. The stemming was performed using a Porter stemmer to eliminate tokens that are meaningless in the given context (Van Rijsbergen, Robertson, & Porter, 1980).

After stemming, the text was passed through a tokenizer for Indo-European languages and then n-gram detection was carried out. Once the initial set of interesting phrases was received, the underlying body of text was examined, and sets of tokens identified that represent a similar body of text. In this case, these multiple tokens were collapsed into a single token.

The following are the distinct steps that were carried out as part of this research:

1. Validating the use of knowledge

2. Knowledge Structure

    a. How are the communities and networks formed?

    b. How the knowledge develops and evolves over a period

c. Ontology of knowledge

d. Evolution of knowledge

3. Formation of cliques within communities

4. Knowledge Adjacency

5. System architecture

3.3.1 Validating the use of knowledge

To validate the importance to knowledge organizations, we conducted a survey across similar organizations based in India. The results of the survey were analyzed using statistical methods to validate the belief that tacit knowledge is important to organizations and they miss it when individuals possessing that knowledge leave the organizations. Chapter 4 provides further details on this aspect of research and results.

3.3.2 Knowledge structure

The knowledge contained within organizations exists in their communication archives and document repositories. Since most of this exists in the form of unstructured data, the first step was to design and provide some structure. A knowledge ontology was defined and algorithms built to create knowledge networks that are suitable for social network analysis. The measures of social network analysis were used to evaluate the evolution of knowledge within groups and individuals. This aspect of the research is further discussed in Chapter 5.

3.3.3 Knowledge adjacency

The communications and documents generated by most knowledge organizations and communities following special taxonomy and vocabulary thus rendering the use of synonyms useless for keyword search. We define phrase similarity based on the probability of two words occurring together. We define this concept as an adjacency which is examined in detail in Chapter 6.

3.3.4　Cliques

Cliques form within any large community and to detect them we use community detection algorithms. We experiment with multiple such algorithms and then choose (Blondel et al., 2008) because of its performance which is detailed in Chapter 6.

3.4　Dataset, tools and compute infrastructure

3.4.1　Dataset

In the context of this research, a dataset that has content in the form of communication was desirable to identify sender and receiver. Content not specific to a particular domain was also required. The best dataset would have been collaboration systems within enterprises, but it is extremely difficult to get access to data stored within these systems. The Usenet was such a system which very closely resembles communities of practice and has content related to many domains. We have used Usenet archive provided by UTZOO, University of Toronto, Department of Zoology as the primary dataset since it consists of meaningful discussions about many topics in that era(Spencer & Wiseman, n.d.).

3.4.2　Parsing

We built a software system that processes the dataset in following steps.

- Parse the mbox format files contained within the archive
- Populate RDBMS (MySQL) from the parsed entities, underlying knowledge networks, important phrases, and other artifacts

Following is the list of software artifacts that were used as part of this research.

- Ubuntu 14.04.3 LTS("Ubuntu," n.d.)
- Wildfly Application Server 8.1.0(WIL, 16AD)
- MySQL 5.5.46 ("MySQL," n.d.)

- MySQL Connector/J 5.1.32 (MySQL, 2006)

- Lingpipe 4.1.0(Alias-i, 2008)

- D3.js (Bostock, 2012)

3.4.3   Network analysis

We parsed the dataset and analyzed using network analysis algorithms. Following sets of software was used for this purpose.

- R 3.2.3 (R Development Core Team, 2008)

- Igraph (Gabor & Tamas, 2006)

3.4.4   Compute infrastructure

The dataset was parsed and analyzed on following set of hardware

- Intel i7-4770K CPU at 3.6 GHz

- 32 GB DDR3 RAM

- Primary HDD 1TB

- Additional NAS Storage 8 TB based on freenas.org

3.4.5   Dataset details

UTZOO Usenet Archive(Spencer & Wiseman, n.d.) consists of 140 tar and gzipped files. These files total up to 2 GB of tar and gzipped data. The binary files have been removed from this data. Once expanded these results in 16 GB of mbox format files. Following are the salient features of the data after processing.

- 237,472 Individuals email addresses mapping to the same number of Individuals

- 1,862,726 distinct Usenet posts

- 1,784,661 distinct edges in the knowledge graph without phrase weight

- 2,115,857 phrases of importance

- 197 GB Total size of the database after processing

3.4.6 Execution time

The algorithms developed and used in this research were not optimized for computational efficiency, the following is the approximate execution times required for each phase of research.

- Parsing messages and storing in the database 45 days
- Parsing database entries to generate time sliced graphml files 35 days

In Chapter 4 we present the results of exploratory research which answers the first of our research question. We look to validate the anecdotal belief that tacit knowledge is really important for the organizations and communities.

CHAPTER 4

4. Importance of Knowledge

Organizations need knowledge for their survival. They spend a significant amount of effort and time to codify and manage the knowledge that they have. But even with all that effort, there is a dimension of knowledge, known as the tacit knowledge that only stays in the heads of the individual who are using it for their day to day work. Organizations need to understand how to access that knowledge best.

4.1 Scale

We wished to establish that organizations value tacit knowledge and use it for the decision making. We did not come across any existing scale that measures the importance and value of tacit knowledge in existing employees and employees who have left the organization.

Organization decision making requires tapping the tacit knowledge of fellow employees (Payne et al., 1993), rationality of any decision making is limited by the time available to make a decision (Simon, 1999) making it easier to reach out to a person rather than searching through the repositories to find the information needed to make the decision. Individuals with tacit knowledge are only known to individuals who are interacting with them (Davenport & Prusak, 2000). Many of the problem solving skills also reside in the minds of individuals in form of tacit knowledge (Grant, 1996b). It is believed that all the tacit knowledge cannot be codified and the presence of tacit knowledge is only known to individuals who are interacting with other individuals who possess it.

To establish the importance of tacit knowledge, we performed exploratory research by surveying the knowledge workers within the organizations by asking questions on whether they felt the need of using tacit knowledge from another individual that is currently not part of their team.

We did not come across any scale that fit our purpose, we created a questionnaire which is appended in Appendix B. The Cronbach alpha for the questionnaire was computed to establish internal consistency for the scale.

Table 1 Reliability Statistics of questionnaire

Cronbach's Alpha	Cronbach's Alpha Based on Standardized Items	N of Items
.867	.852	16

The Cronbach alpha for our questionnaire is 0.852 which indicates a high level of internal consistency for out scale.

Table 2 Statistics for each question in questionnaire

	Scale Mean if Item Deleted	Scale Variance if Item Deleted	Corrected Item-Total Correlation	Squared Multiple Correlation	Cronbach's Alpha if Item Deleted
emp_categ	15.26	230.525	.004	.040	.875
role_categ	16.12	238.559	-.359	.963	.879
are_you_org_sys_proc_expert	17.23	228.300	.247	.311	.869
contact_org_sme	16.01	172.365	.877	.963	.836
contact_sme_left	16.77	192.745	.784	.902	.846
peers_need_sys_proc_help	16.00	164.979	.848	.927	.839
peers_need_sys_proc_helped_by_team	16.25	172.353	.840	.912	.839
staff_sys_proc_expert	17.86	228.064	.340	.940	.869
others_request_your_help_for_sys_org_proc	17.37	215.669	.265	.946	.871
when_others_request_help_your_team_members_can_help	17.45	217.545	.251	.878	.871
are_you_sme	17.75	222.149	.722	.902	.864
how_often_you_need_other_group_sme	16.73	189.014	.655	.977	.852
how_often_you_need_sme_who_left	17.17	201.370	.648	.923	.854
sme_in_your_team	17.80	223.664	.663	.907	.865
how_often_others_need_your_sme	16.99	190.683	.625	.903	.853
sme_replaced_by_doc_videos	17.16	198.354	.607	.893	.855

We look at the last column, *Cronbach's Alpha if Item Deleted*, for 10 questions the Cronbach's Alpha would reduce if we delete that item. We can also look at *Corrected Item-Total Correlation*, and conclude that 12 questions have a value higher than 0.3. The question that is most troublesome is the first and second questions, employee category and role category. We dropped those questions and did not use them for any further processing.

4.2   Model

This model is designed from the knowledge seeker's point of view. If somebody is looking for tacit knowledge in a particular field, how is it sought?

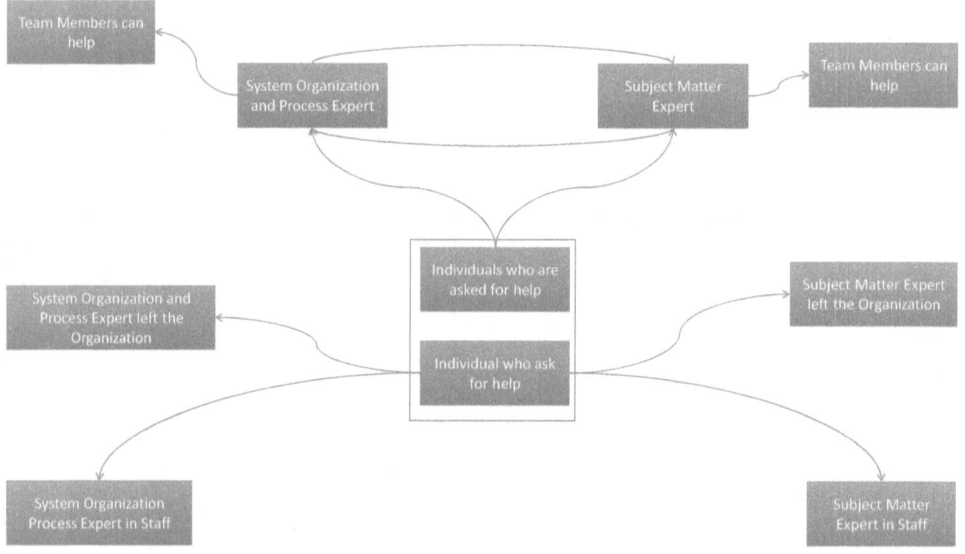

Figure 13 Seeker's knowledge transfer model; relationships of individuals who ask for help and who are asked for help

It is believed that tacit knowledge cannot be created(Polanyi, 1966) but it can be transferred through shared social experiences (Ikujiro Nonaka, 1994), which he called Socialization. We are using Nonaka (1994) definition of tacit knowledge as far as transmission of knowledge is concerned. It is because of this definition we want to evaluate if the tacit knowledge of individual is diffused within his team members.

68

There is general agreement that tacit knowledge disappears when any individual leaves the organization. In this research, the objective is to validate that tacit knowledge exists, it is valuable, and there is a net loss to the organization when some individual leaves. It is also established that tacit knowledge existing in the organization is concentrated in a group of different individuals with varying degrees of expertise in any subject.

If the knowledge can be codified, then the best approach is to codify and share it using the organization's knowledge management system. What is of interest in this research is the knowledge that cannot be codified. As regards Nonaka's model of knowledge creation and transfer, as shown in Figure 9, the primary concern of this research is with the left half of the matrix of socialization and internalization. The best way to access tacit knowledge is to find the individuals who possess it.

The initial model in Figure 13 explains the relationships between hypotheses. The following are the salient points of the model:

- Individuals can ask for help from other individuals
- Some individuals are subject matter experts
- Some individuals are system organization and process experts
- Individuals may be managers and have staff members who are SMEs or SOPEs.
- Individuals who are experts leave organizations and need to be contacted.

4.3 Methodology

To understand the importance of tacit knowledge in the eyes of knowledge workers, we surveyed knowledge organizations based in India. The survey was targeted primarily at an Indian audience, but since it was administered over the internet, there was no way to guarantee that other nationalities did not take part. It was disseminated through email groups, LinkedIn groups and other social media thus it is difficult to conclude how many individuals received it.

Total 50 emails were also sent directly to individuals in addition to posts to groups. In total, 146 responses were received. Appendix B describes the survey design in detail.

One justifiable use of a convenience sample is for exploratory purposes, that is, to get different views on the dimensions of a problem, to probe for possible explanations or hypotheses, and to explore constructs for dealing with particular problems or issues(Ferber, 1977).

The population for purpose of this research was all knowledge workers working within different corporations and in different communities of practice. The survey was disseminated to individuals with background in medicine, engineering and management. These individuals were then requested to forward the survey to their co-workers. No specific restriction was placed on types of companies to which survey was forwarded. The medium chosen for disseminating survey were best suited to reach the target audience. Due to exploratory nature of research and time constraints, we could not reach out to an exhaustive list of corporations and communities of practice. Since the selection of sample was not random and a large percentage of samples were from Information Technology companies, there could be bias in the sample and it may not be representative of population. Therefore, the findings of this research cannot be generalized across all types of corporations and communities of practice.

4.4 Design of construct

The tacit knowledge of the individual needs to be measured. However, this is a challenging task. In knowledge organizations, individuals will reach out to their colleagues and team members when they are searching for knowledge. It is almost second nature. This behavior of individuals and teams is used as a proxy of what knowledge individuals and teams in the organizations possess. If an individual is known to be an expert within an organization, it is assumed that he or she possesses some tacit knowledge about the topic. Most of the knowledge-organizations have two sets of experts who were tapped for tacit knowledge. These are subject matter experts and organization and process experts. It was also necessary to measure whether tacit knowledge diffuses, i.e., if an individual possesses some knowledge, it will diffuse to his or her closest team members. Specifically, it was also measured whether people even reach out to others who might have left the organization. Based on the combinations of the above criteria, the variables were determined that need to be measured and that helped with the design of the questionnaire.

4.5 Hypothesis

The following are the set of hypotheses that need to be examined.

4.5.1 SOP experts soliciting help from subject matter experts

An individual's propensity to solicit help from subject matter experts from the same organization or from the SMEs who have left does not depend on whether they are organization system and process experts.

H_0 *Mean of are_you_org_sys_proc_expert is same as the mean of contact_org_sme.*
H_a *Mean of are_you_org_sysproc_expert is NOT same as the mean of contact_org_sme.*
H_0 *Mean of are_you_org_sys_proc_expert is same as mean of contact_sme_left*
H_a *Mean of are_you_org_sys_proc_expert is NOT same as mean of contact_sme_left*

Since are_you_org_sys_proc_expert is a categorical variable, we perform ANOVA. While testing for contact_org_sme, there was the homogeneity of variances, as assessed by Levene's test for equality of variances ($p = .065$). There were no statistically significant differences in individuals' propensity to contact their subject matter experts irrespective of them being a system and process experts or not. $F(1,144) = 1.931, p = .167$.

Similarly, the test for contact_sme_left showed, there was the homogeneity of variances, as assessed by Levene's test for equality of variances ($p = .928$). There were no statistically significant differences in individuals' propensity to contact subject matter experts who left the organization irrespective of them being a system and process experts or not. $F(1,144) = .448, p = .504$.

4.5.2 Subject matter experts soliciting help from subject matter experts

An individual's propensity to solicit help from subject matter experts from the same organization or from the SMEs who have left does not depend on the fact whether they are organization system and process experts.

H_0 *Mean of are_you_sme is same as mean of the contact_org_sme.*
H_a *Mean of are_you_sme is NOT same as the mean of contact_org_sme.*
H_0 *Mean of are_you_sme is same as the mean of contact_sme_left*
H_a *Mean of are_you_sme is NOT same as the mean of contact_sme_left*

Since *are_you_sme* is a categorical variable, we perform ANOVA between *are_you_sme* and other two variables. Following are the results.

While testing for *contact_org_sme*, the assumption of homogeneity of variances was violated as assessed by Levene's test for equality of variances (p = .000). There were statistically significant differences in individuals' propensity to contact their subject matter experts depending on them being a subject matter expert or not. $F(1,144) = 69.184, p < .0005$. Individuals were more likely to contact their subject matter experts if they themselves were subject matter experts.

While testing for *contact_sme_left*, there was the homogeneity of variances, as assessed by Levene's test for equality of variances (p = .297). There were statistically significant differences in individuals' propensity to contact their subject matter experts depending on them being a subject matter expert or not. $F(1,144) = 64.381, p < .0005$. Individuals were more likely to contact subject matter experts who left their groups or companies if they themselves were subject matter experts.

Next we looked at the strength of the relationship. For *contact_org_sme* it is as below.

$$\eta^2 = \frac{SS_{between}}{SS_{total}} = \frac{257.016}{791.973} = .324$$

Above computation tells us that 32.45% of the variation in *contact_org_sme* is because of *are_you_sme*. Next, we looked at the strength of the relationship. For *contact_sme_left* it is as below.

$$\eta^2 = \frac{SS_{between}}{SS_{total}} = \frac{123.441}{399.541} = .308$$

The result tells us that 30.89% of the variation in *contact_sme_left* is because of *are_you_sme*. As we can see from the results, there is a significant correlation between people who are subject matter expert and their tendency to solicit help from SMEs. At a confidence level of 99%, we reject both the hypothesis and conclude that subject matter experts have higher tendency to solicit help from other subject matter experts within the organization as well as from the SMEs who have left the organization.

4.5.3 Individuals who have SMEs in their staff

Next, we want to evaluate if individual's having SMEs in their staff impact their propensity to solicit help.

H_0 *Mean of sme_in_your_team is same as the mean of contact_org_sme.*
H_a *Mean of sme_in_your_team is NOT same as the mean of contact_org_sme.*
H_0 *Mean of sme_in_your_team is same as the mean of contact_ sme_left*
H_a *Mean of sme_in_your_team is NOT same as the mean of contact_sme_left*

We look at the coefficient of correlation between these variables and here are the results.

While testing for *contact_org_sme*, the assumption of homogeneity of variances was violated as assessed by Levene's test for equality of variances (p = .000). There were statistically significant differences in individuals' propensity to contact their subject matter experts depending on them being a subject matter expert or not. $F(1,144) = 45.107, p < .0005$.

Individuals were more likely to contact their subject matter experts if they had subject matter experts as team members.

While testing for *contact_sme_left*, there was the homogeneity of variances, as assessed by Levene's test for equality of variances (p = .983). There were statistically significant differences in individuals' propensity to contact their subject matter experts depending on them being a subject matter expert or not. $F(1,144) = 45.700, p < .0005$. Individuals were more likely to contact subject matter experts who left their groups or companies if they themselves were subject matter experts.

Next, we looked at the strength of the relationship. For *contact_org_sme* it is as below.

$$\eta^2 = \frac{SS_{between}}{SS_{total}} = \frac{188.908}{791.973} = .2385$$

The result tells us that 23.85% of the variation in *contact_org_sme* is because of *are_you_sme*.

Next, we looked at the strength of the relationship. For *contact_sme_left* it is as below.

$$\eta^2 = \frac{SS_{between}}{SS_{total}} = \frac{96.252}{399.541} = .2409$$

The result tells us that 24.09% of the variation in *contact_sme_left* is because of *are_you_sme*.

As we can see from the results, there is a significant relationship between people who have SMEs in their team and their tendency to solicit help from SMEs. We reject both the hypothesis and conclude that teams with SMEs in them have higher tendency to solicit help from other subject matter experts within the organization as well as from the SMEs who have left the organization.

4.5.4 Choosing appropriate Correlation Coefficient

For all the variables involved in the hypotheses that are enumerated in 4.5.5, 4.5.6, 4.5.7, and 4.5.8 we can look at Shapiro-Wilk statistics for each of them. Table 3 contains the results of test of normality for different variables.

Table 3 Test of Normality; all the variables fail the test of normality

	Kolmogorov-Smirnov[a]			Shapiro-Wilk		
	Statistic	df	Sig.	Statistic	df	Sig.
contact_org_sme	.350	146	.000	.755	146	.000
contact_sme_left	.318	146	.000	.748	146	.000
peers_need_sys_proc_help	.398	146	.000	.689	146	.000
peers_need_sys_proc_helped_by_team	.395	146	.000	.703	146	.000
staff_sys_proc_expert	.520	146	.000	.396	146	.000
others_request_your_help_for_sys_org_proc	.514	146	.000	.410	146	.000
when_others_request_help_your_team_members_can_help	.520	146	.000	.390	146	.000
are_you_sme	.469	146	.000	.535	146	.000
how_often_you_need_other_group_sme	.437	146	.000	.627	146	.000
how_often_you_need_sme_who_left	.418	146	.000	.603	146	.000
sme_in_your_team	.495	146	.000	.480	146	.000
how_often_others_need_your_sme	.483	146	.000	.518	146	.000
sme_replaced_by_doc_videos	.483	146	.000	.523	146	.000

a. Lilliefors Significance Correction

A survey of literature on choice of appropriate correlation coefficient comes up with conflicting results. Normality assumption underlying r is robust indicating that violation of the population normality assumption does not seriously alter the interpretation of r and violation of the population normality assumption appears, therefore, to be insufficient reason to deny r a place as a major tool(Zeller & Levine, 1974). Pearson r is insensitive to rather extreme violations of

75

the basic assumptions of normality and type of measurement scale. Failure to meet the basic assumptions separately or in combinations had little effect upon the obtained distributions of r (Havlicek & Peterson, 1976).

Distribution of Pearson's correlation may be quite sensitive to non-normality and correlation analyses should be limited to situations where underlying data is very close to normal(Kowalski, 1972). All of the variables are not normal. Spearman's rank correlation coefficient is more appropriate when one or both variables are skewed or ordinal and is robust when extreme values are present(Mukaka, 2012). Confidence intervals for Spearman's r_s are less reliable and less interpretable than confidence interval for Kendall's τ (Kendall & Gibbons, 1990). In the light of these conflicting conclusions regarding assumptions related to different correlation coefficients, all three correlation coefficients were computed and analyzed for the results.

Rest of the hypotheses have ordinal variable, so the correlation coefficient was computed.

4.5.5 Individuals who are asked for help

Next, we want to evaluate whether a group/individual being asked for help impacts his propensity to seek help from other SMEs. There are three variables can be used as a proxy for people who are asked for help. These are *peers_need_sys_proc_help*, *how_often_others_need_your_sme*.

H_0 Mean of *peers_need_sys_proc_help* is same as the mean of *contact_org_sme* .
H_a Mean of *peers_need_sys_proc_help* is NOT same as the mean of *contact_org_sme*.
H_0 Mean of *peers_need_sys_proc_help* is same as the mean of *contact_sme_left*
H_a Mean of *peers_need_sys_proc_help* is NOT same as the mean of *contact_sme_left*
H_0 Mean of *how_often_others_need_your_sme* is same as the mean of *contact_org_sme*.
H_a The mean of *how_often_others_need_your_sme* is NOT same as the mean of *contact_org_sme*.
H_0 Mean of *how_often_others_need_your_sme* is same as the mean of *contact_sme_left*
H_a Mean of *how_often_others_need_your_sme* is NOT same as the mean of *contact_sme_left*

The test of normality on all the above variables assessed by Shapiro-Wilk's test shows that the variables are not normally distributed (p<0.05).

There was a strong positive correlation between groups who look for system and process help and their inclination to look for subject matter expert help. $r = .792, p < .0005$; $r_s = .803, p < .0005$; $\tau = .712, p < .0005$. Detailed results are enumerated in Appendix F.

There was a positive correlation between groups who look for system and process help and their inclination to look for help from subject matter experts who have left the organization. $r = .640, p < .0005$; $r_s = .758, p < .0005$; $\tau = .652, p < .0005$.

There was a positive correlation between groups whose subject matter expert's help is solicited by other groups and them asking for help from organization subject matter experts. $r = .505, p < .0005$; $r_s = .523, p < .0005$; $\tau = .473, p < .0005$.

There was a positive correlation between groups whose subject matter expert's help is solicited by other groups and them asking for help from subject matter experts who have left the organization. $r = .507, p < .0005$; $r_s = .551, p < .0005$; $\tau = .495, p < .0005$. Following table summarizes all the results.

Table 4 Correlation between for *contact_org_sme* and *peers_need_sys_proc_help*

	peers_need_sys_proc_help
Pearson's Correlation	$r = .792, p < .0005$
Kendall's Tau B	$\tau = .712, p < .0005$
Spearman's Rho	$r_s = .803, p < .0005$

Table 5 Correlation between *contact_org_sme* and *how_often_others_need_your_sme*

	how_often_others_need_your_sme
Pearson's Correlation	$r = .505, p < .0005$
Kendall's Tau B	$\tau = .473, p < .0005$
Spearman's Rho	$r_s = .523, p < .0005$

Table 6 Correlation between *contact_sme_left* and *peers_need_sys_proc_help*

	peers_need_sys_proc_help
Pearson's Correlation	$r = .640, p < .0005$
Kendall's Tau B	$\tau = .652, p < .0005$
Spearman's Rho	$r_s = .758, p < .0005$

Table 7 Correlation between *contact_sme_left* and *how_often_others_need_your_sme*

	how_often_others_need_your_sme
Pearson's Correlation	$r = .507, p < .0005$
Kendall's Tau B	$\tau = .495, p < .0005$
Spearman's Rho	$r_s = .551, p < .0005$

As we can see from the results, there is a significant correlation between individuals who are asked help and their tendency to solicit help from SMEs and individuals who have team members who are asked help and their tendency to solicit help from SMEs. We reject both the hypothesis and conclude that individuals who are asked for help or individuals who are part of teams with team members who are asked help have higher tendency to solicit help from other subject matter experts within the organization as well as from the SMEs who have left the organization.

4.5.6 Individuals who solicit help

Next, we want to evaluate if the individual who solicit help from other groups within the same organization also solicit help from ex-employees.

H_0 *Mean of contact_org_sme is same as the mean of contact_sme_left.*
H_a *Mean of contact_org_sme is NOT same as the mean of contact_sme_left.*

We look at the coefficient of correlation between these variables and here are the results. The test of normality on all the above variables assessed by Shapiro-Wilk's test shows that the variables are not normally distributed (p<0.05).

There was a strong positive correlation between groups who take the help of subject matter expert within their organization and their propensity to take help from subject matter experts who have left the organization $r = .824, p < .0005$; $r_s = .924, p < .0005$; $\tau = .813, p < .0005$. Detailed results are enumerated in Appendix F.

Table 8 Correlation between *contact_org_sme and contact_sme_left*

Pearson's Correlation	$r = .824, p < .0005$
Kendall's Tau B	$\tau = .813, p < .0005$
Spearman's Rho	$r_s = .924, p < .0005$

As we can see from the results, there is significant very high positive correlation between people who have taken help from SMEs from other groups in organization and people who have taken help from SMEs who have left.

4.5.7 Giving help vs. taking help

Next, we want to evaluate if the individual who receives a request for help from other group is more likely to ask for help.

H_0 *Mean of how_often_others_need_your_sme is same as mean of contact_org_sme.*
H_a *The mean of how_often_others_need_your_sme is NOT same as the mean of contact_org_sme.*
We look at the coefficient of correlation between these variables and here are the results. The test of normality on all the above variables assessed by Shapiro-Wilk's test shows that the variables are not normally distributed (p<0.05).

There was a positive correlation between groups having a sought after subject matter experts and their propensity to take help of subject matter expert within their organization. $r = .505, p < .0005$; $r_s = .523, p < .0005$; $\tau = .473, p < .0005$. Detailed results are enumerated in Appendix F.

Table 9 Correlation between *contact_org_sme and how_often_others_need_your_sme*

Pearson's Correlation	$r = .505, p < .0005$
Kendall's Tau B	$\tau = .473, p < .0005$
Spearman's Rho	$r_s = .505, p < .0005$

As we can see from the results, there is a positive and significant correlation between individuals who were asked for help and who ask for help from others.

4.5.8 Spreading expertise to group

Next, we want to evaluate if an individual's expertise also percolates down to his team members. We asked in the survey how many times team members can satisfy the needs of somebody who solicited for help.

H_0 *Mean of when_others_request_help_your_team_members_can_help is same as mean of others_request_your_help_for_sys_org_proc.*

H_a *Mean of when_others_request_help_your_team_members_can_help is NOT same as mean of others_request_your_help_for_sys_org_proc.*

The test of normality on all the above variables assessed by Shapiro-Wilk's test shows that the variables are not normally distributed (p<0.05).

There was a strong positive correlation between groups having a sought after subject matter experts and their propensity to take help of subject matter expert within their organization. $r = .932, p < .0005$; $r_s = .943, p < .0005$; $\tau = .920, p < .0005$. Detailed results are enumerated in Appendix F.

Table 10 Correlation between *others_request_your_help_for_sys_org_proc and when_others_request_help_your_team_members_can_help*

Pearson's Correlation	$r = .932, p < .0005$
Kendall's Tau B	$\tau = .920, p < .0005$
Spearman's Rho	$r_s = .943, p < .0005$

The results show that there is a positive and very high correlation between individuals who are asked for help and the cases where their team members can provide that help (Avasthi, Vinay & Dey, 2015).

4.6 Conclusions

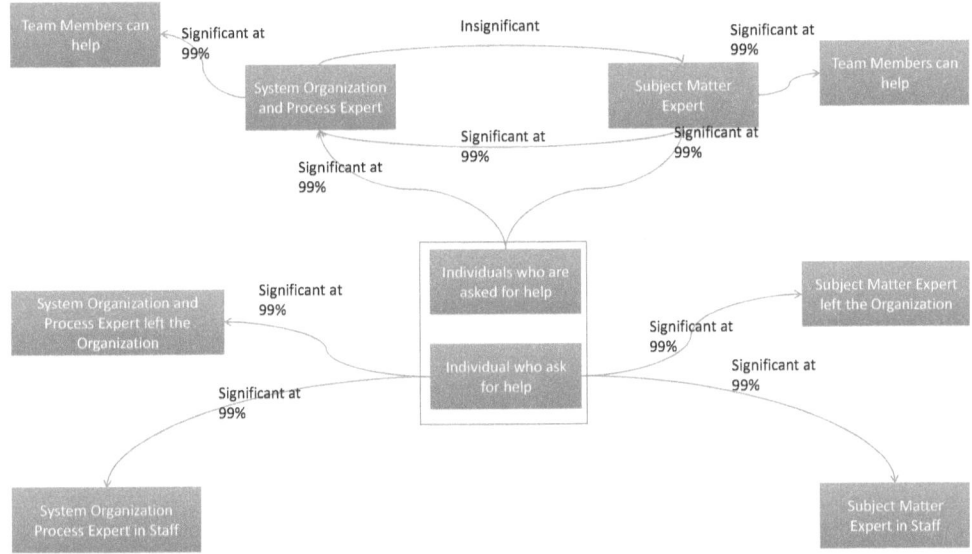

Figure 14 Seeker's knowledge transfer model with relationship significance populated

As part of this exercise, we wanted to evaluate some of the well-accepted beliefs that tacit knowledge leaves with an individual when they leave the organization. It is also believed that team members who worked with that individual will have expertise in adjacent tacit knowledge.

Except for subject matter experts, all of the accepted beliefs seem to hold true.

There is no pattern of knowledge giver and knowledge receiver. Individuals who provide most help also seek the most amount of help. Teams that have experts within their teams that others solicit, also value them more and ask more from experts outside. Here is the final model modified with the results.

Figure 14 summarizes the results of this work which is as below.

- System and process experts Do not seem to solicit help from subject matter experts whether in the organizations or those who have left the organization.

- Individuals who are subject matter expert ask for help from their ex-colleagues (whether within an organization or individuals that have left the organization). The conclusion signifies the fact that they believe that some knowledge is carried by them and the best way to access that knowledge is to get in touch with those individuals.
- We also see that individuals who are most likely to ask help from individuals within the organization would also reach out to ex-colleagues who have left the group or organization.

We have established that tacit knowledge is important for organizations, groups, and individuals. They miss it when individuals with tacit knowledge leave the organization. Since one way to get access to tacit knowledge is Socialization, we adopt social network concept in the form of knowledge networks and discuss a methodology to reach individuals with such tacit knowledge by traversing through knowledge networks. Chapter 5 describes methodology and algorithms that are used to build these knowledge networks from the communications across individuals within the organizations.

CHAPTER 5

5. Knowledge Networks and Evolution

5.1 Method

We organized the communication archive into a knowledge network, and then this knowledge network was used to answer other questions. We constructed knowledge networks for a period and generated the network measures like betweenness centrality, indegree, and outdegree for all the vertices. These changes in the measures were used to compute changes in the topic importance, individual expertise.

As the knowledge networks were evolving, we used Louvain method (Blondel et al., 2008) for identifying cliques within the network.

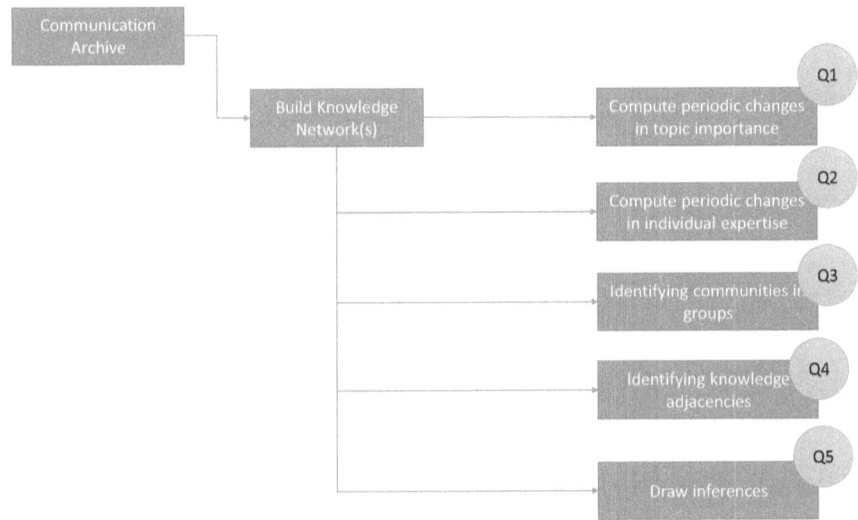

Figure 15 Our research approach; building knowledge networks from dataset and using those networks for further computation

As the topic importance evolves, we computed the topic adjacencies, and that were used to estimate tacit knowledge expertise. Figure 15 describes our research approach to the problem under consideration. The approach maps each of the research questions in terms of their contribution.

5.1.1 Network Creation

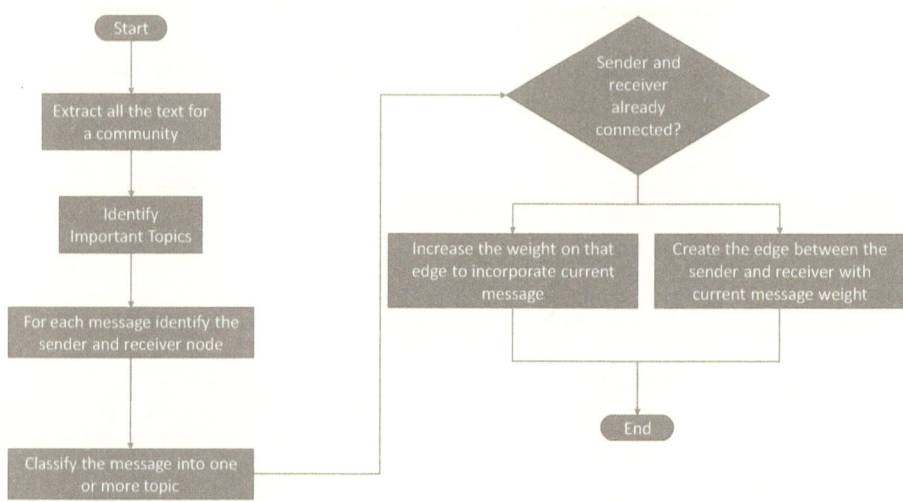

Figure 16 Knowledge network generation flow chart

Usenet is a collection of multiple communities of practice, and we used its archives as a repository of communication archives of those communities of practice. The UTZOO Usenet Archive(Spencer & Wiseman, n.d.) is used to analyze the communication patterns across individuals and building of knowledge networks and graphs. Since the archive consists of Usenet messages, there is no strict sender-receiver relationship. We look at each message and the message in whose response it was sent, and then we create a sender-receiver relationship.

We generate directed graphs from these messages with one weighted edge between two nodes per topic. Once the graphs are generated, they are analyzed using network theory and inferences are drawn. Figure 16 defines the detailed algorithm of how the graphs are generated from the archive.

Following steps are followed to generate the graphs.

- Identify the important topics

- For each message repeat the following

    - Classify the message in topics

    - If the sender and receiver are already connected, then increase the weights to account for the current message

    - If the sender and receiver are not connected, then create a new connection with the weight of the current message

We ran this algorithm on a moving window of messages each spanning six months. The results in a set of networks that incorporate the change in networks over a period.

5.1.2   Topic identification

Text mining techniques were used to extract interesting phrases, the similarity of texts to enrich the retrieval mechanisms. Statistical and machine learning techniques were used to identify and build patterns in data over time.

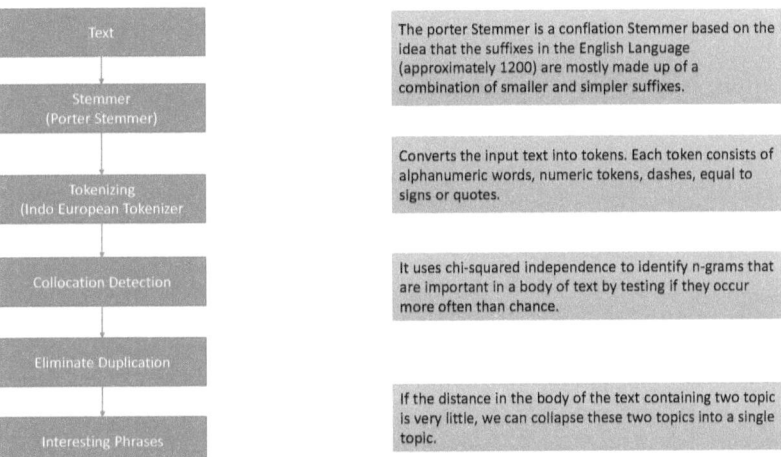

Figure 17 How do we identify topics

Figure 17 describes the methodology for identifying phrases. The stemming was performed using a Porter stemmer to eliminate tokens that are meaningless(Bishop, 2006). After stemming, the text was passed through a tokenizer for Indo-European languages and then n-gram detection is done. Once we had our initial set of interesting phrases, we evaluated the underlying body of text for the source from where the tokens came for similarity to identify a duplicate representation of the same token, and we collapsed these multiple tokens into a single token.

5.1.3 Knowledge Networks

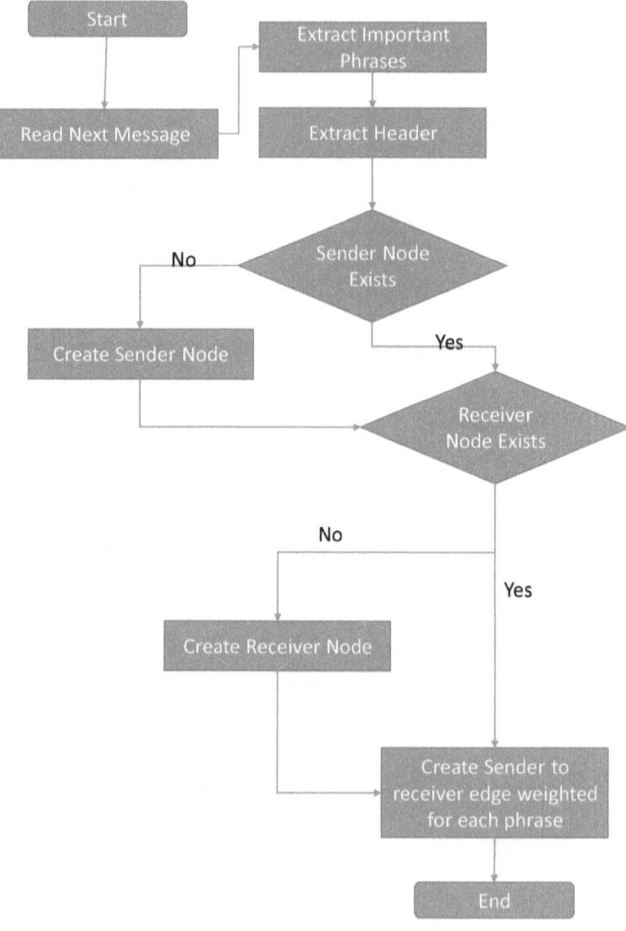

Figure 18 Processing a message for knowledge network

Knowledge networks are networks that are formed between entities whether communities, individuals, departments or organizations because of exchange of knowledge across them(M. Hansen, 2002). Such networks are extremely useful in identifying needed expertise and information that somebody may desire. Every individual seeking knowledge defines his process of discovery. Such discovery aims to find already existing artifacts related to the knowledge that the individual is looking for or find the individual that has expertise in that knowledge and is accessible to the individual looking for knowledge. The current state of the art in knowledge discovery brings together technologies like statistics, machine learning, and artificial intelligence.

There has been an attempt to model knowledge in a database and then use data mining techniques as a tool for knowledge discovery(Fayyad, Piatetsky-Shapiro, & Smyth, 1996). These solutions require significant effort on the part of knowledge creators to convert the knowledge into a format that can be useful for such system.

5.1.3.1 The algorithm

The unit of processing is a single message. As the system comes in the knowledge of single messages, the message is processed, and that results in changes to already existing knowledge network. Figure 18 describes the algorithm that is run each time a new message is received. The message is parsed, header and body are extracted, and each message processing results in either addition of one or more edges in the network or increase in weights of one or more edges.

5.1.4 Applicability of past networks

The following question arises; how relevant past knowledge networks are in today's times? The World has changed, technology has changed, do we even consider knowledge creation and sharing mechanisms prevalent more than two decades ago relevant today? To understand that we tried to see if there has been any change in the method of evolution of knowledge networks over the years. We looked at the knowledge graph measures over two separate windows of half years. We can observe, the basic characteristics of the network Do not change. Individuals who attain higher centrality, indegree and outdegree stay on the top.

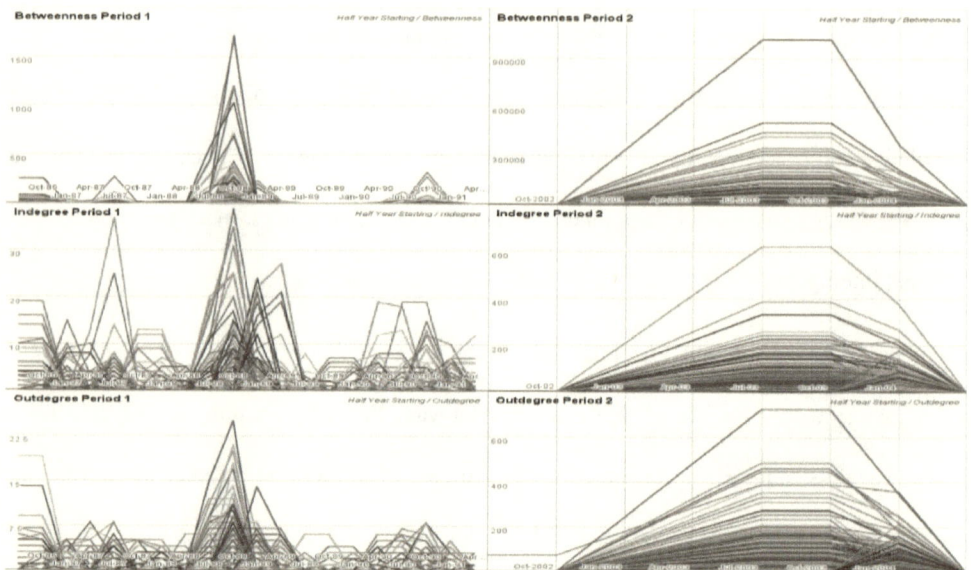

Figure 19 Knowledge graph measures over the years, high betweenness, indegree and outdegree individuals keep their place

5.2 Visualizing a network

Even in most cohesive CoPs, cliques and clusters take shape. Within the context of a large body of knowledge, different individuals are interested in different sub-topics, and that causes cliques to from within the communities. It is important to identify these groups because they point towards an important sub-topic within the larger body of knowledge in that community of practice.

Figure 20 shows the knowledge network of the sci.physics community during the period April-1990 to October 1990. The network points us to interesting characteristics.

The network consists of around seven distinct clusters. These are completely unconnected clusters

- There are three distinct nodes with a very large concentration of incoming edges. These are marked in black.
- There are a fair number of nodes with a large number of outgoing edges.

The width of the edge denotes the density of communication for that topic.

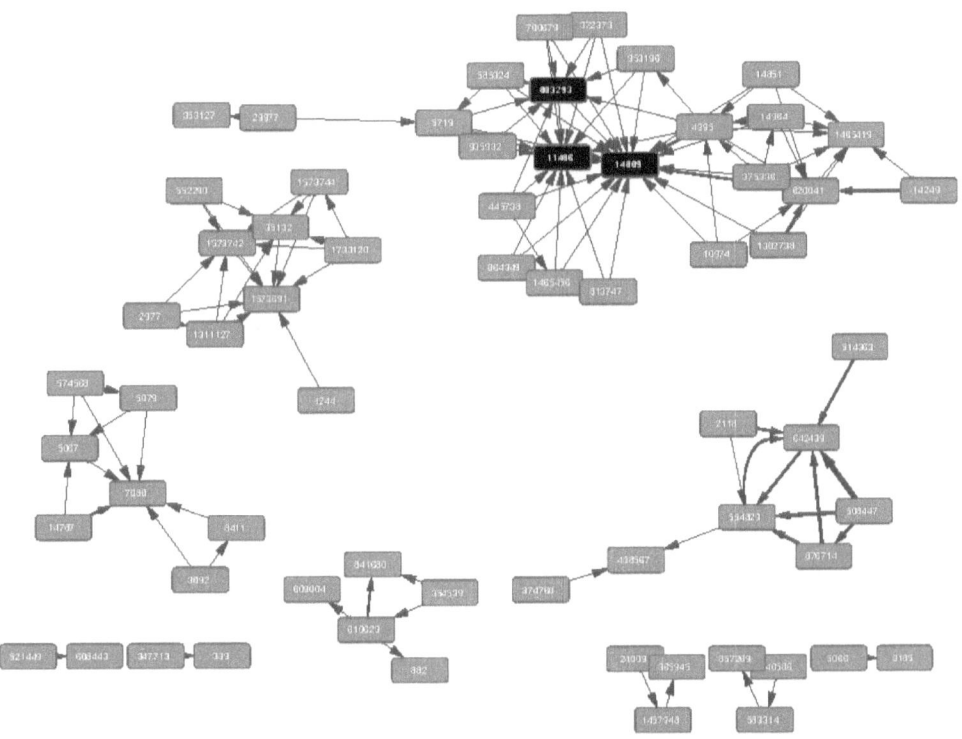

Figure 20 Typical community network during a period with clustered sub-networks

5.3 Knowledge evolution

The next question to be addressed concerns the evolution of knowledge in communities. We have observed that while there are central concepts (denoted by phrases) which are always discussed in a community, there are other concepts that change over the years. Figure 21 depicts a typical knowledge network over a period of half a year.

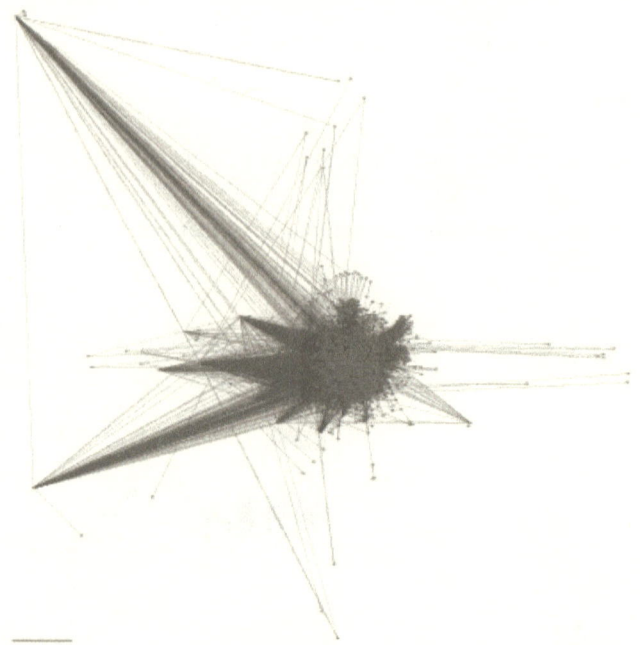

Figure 21 Knowledge network of sci.physics for Oct-2003 second half

5.3.1 Evolution of communities

5.3.1.1 *Out neighborhood distribution*

Figure 22 Out-neighborhood connectivity across nodes reduces with increase in neighbors

The following chart in Figure 22 shows the neighborhood connectivity distribution taking only out edges into account. As expected, nodes with the higher number will have lower average connectivity because most of its neighbors will not have very high connectivity and thus the average will be lower.

5.3.1.2 Betweenness centrality

Figure 23 shows the betweenness centrality for the above network. In most cases, a more connected individual will show a higher betweenness centrality.

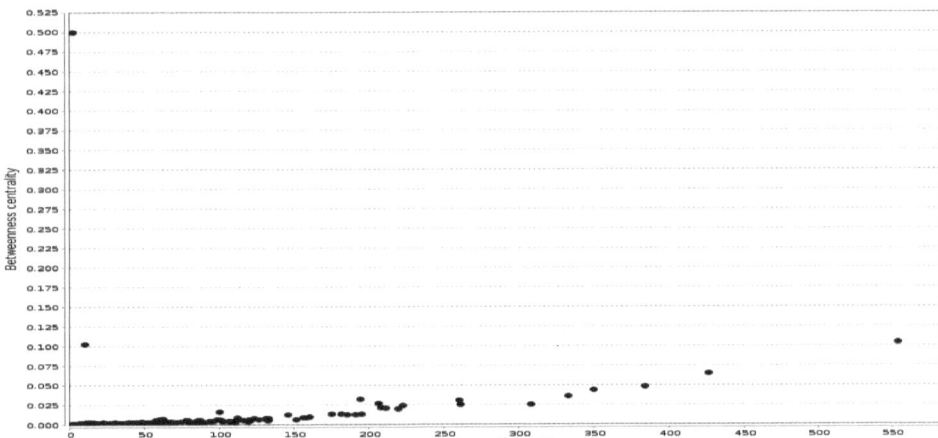

Figure 23 Betweenness centrality increases in nodes as number of neighbors increases

5.3.1.3 Important phrases in a community

We looked at the important topics being discussed within the communities and how these evolve over a period. As we can see in Figure 24, the communities, over a period change their primary interests. There is no gradual increase and decrease in importance. Many phrases which are important at one point in time, become non-important as the time passes.

5.3.1.4 Out-degree distribution

Following chart in Figure 25 shows the out-degree distribution of the graph. As expected, very few individuals in a community will show a high out-degree. Most members of a community are silent participants.

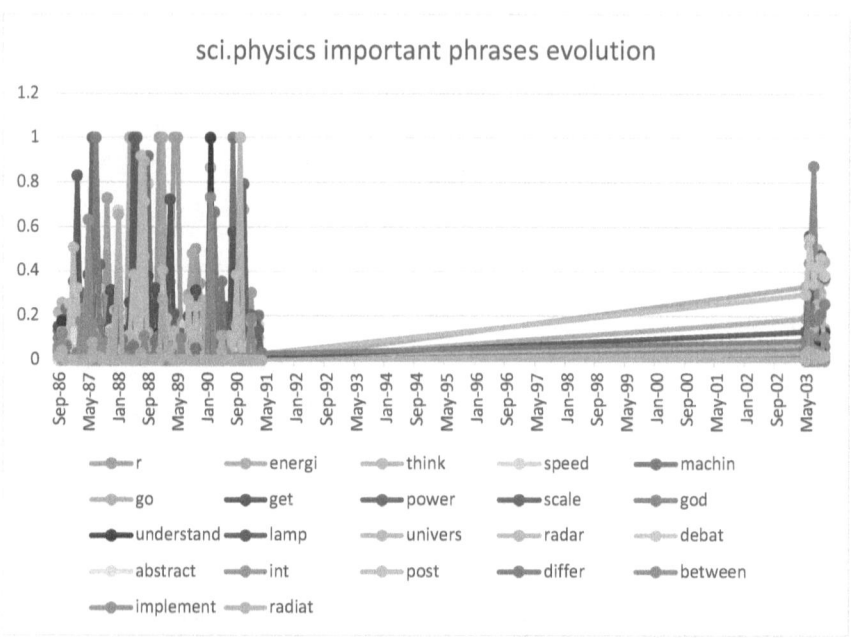

Figure 24 Important phrases over the years in community sci.physics, important phrases change over the time

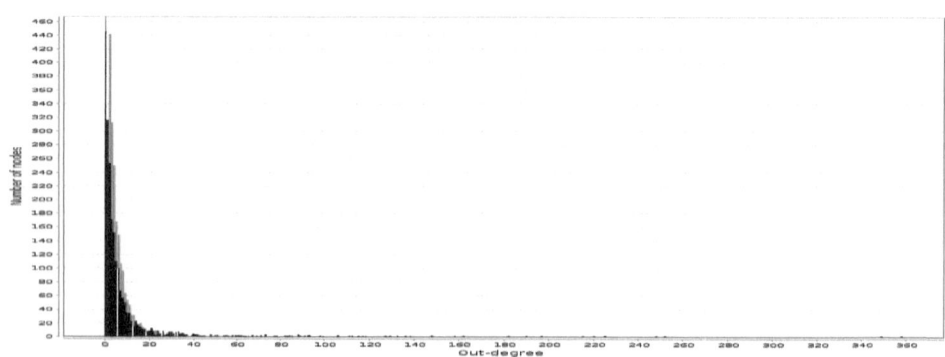

Figure 25 Out degree distribution of the network for community

5.4 Evolution of an individual

We also look at how an individual evolves within a community. We can observe from Figure 26, Figure 27, and Figure 28 as individuals evolve within a community; their position does not easily diminish over a period. An individual can rise to importance within a community, but generally, he does not diminish over time. We examine the role of an individual within the community and evolution of this role over a period.

5.4.1 Betweenness

Figure 26 depicts the evolution of an individual's betweenness over a period. The betweenness of an individual is the measure of the individual's centrality in the network. It is derived from the number of shortest paths between any two nodes which pass through that node. Therefore, higher betweenness signifies higher incidence of an individual acting as a go-to contact for any two individuals. Such an individual is a matchmaker in a community. He or she may not be the source of knowledge himself but knows who possesses it.

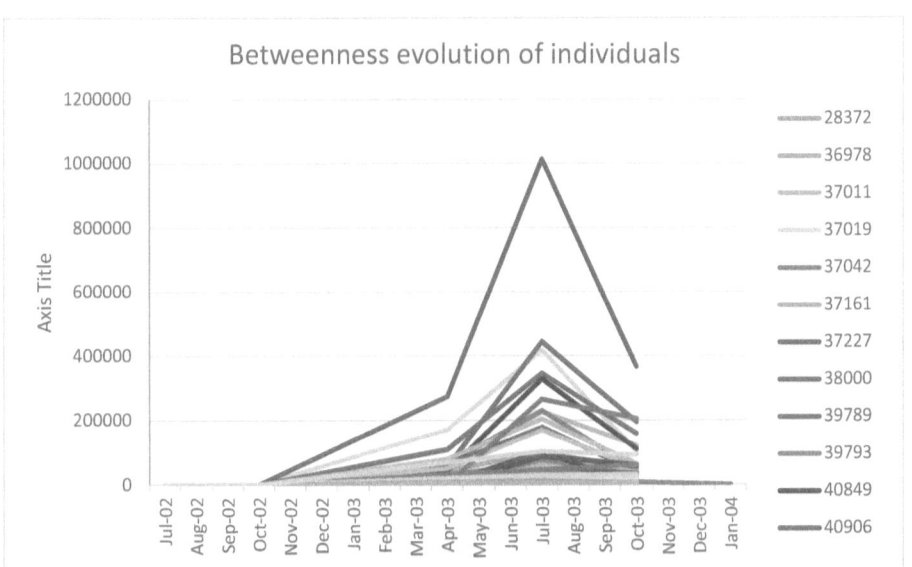

Figure 26 Betweenness centrality evolution of individual over the years; individuals with higher score retain their position

5.4.2 Indegree

Figure 27 depicts the evolution of an individual's indegree over a period. The indegree of an individual is derived from the number of incoming edges to a node. A node in the knowledge graph represents an individual. Therefore, higher indegree signifies the higher number of communication where the individual is a receiver.

In conceptual terms, we could define this as how sought after an individual is within a community. He could be the individual who is asking most questions, and most experts seem to be responsive to such requests by providing an answer. He or she could also be an individual who triggers off the most interesting conversations.

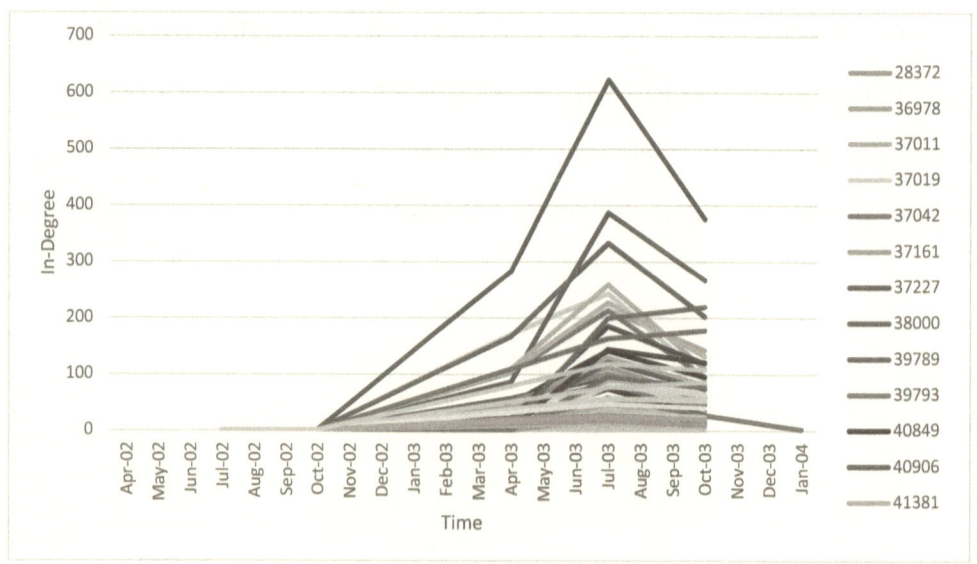

Figure 27 In degree evolution of an individual over the years; individuals with higher score retain their position

5.4.3 Outdegree

The outdegree of an individual is derived from the number of outgoing edges from a node. Thus, higher outdegree signifies the higher number of communications where the individual is a sender. In conceptual terms, we could define this as a measure of how prolific an individual is within a community. Figure 28 depicts the evolution of an individual's outdegree over a period.

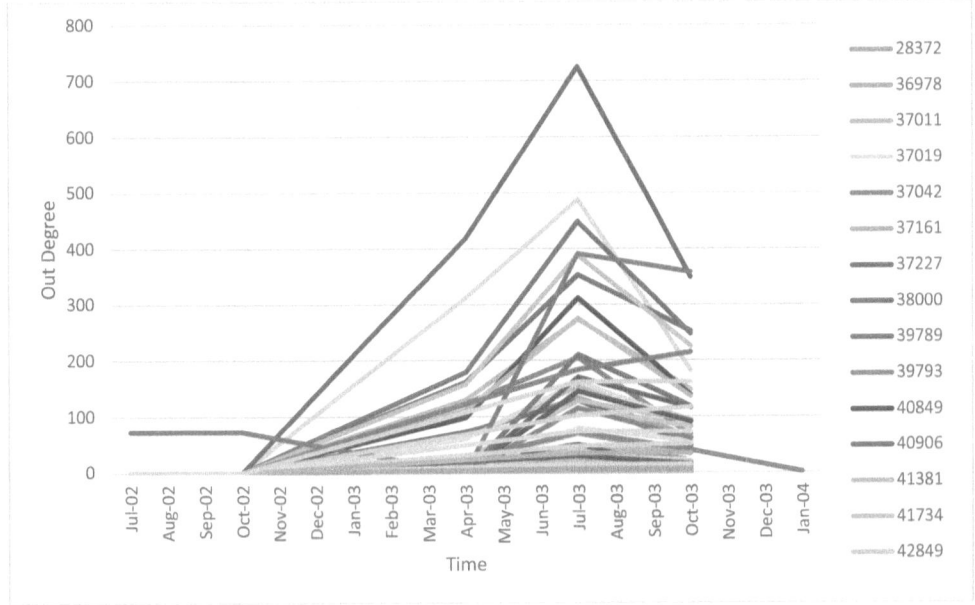

Figure 28 Out Degree evolution of an individual over the years; individuals with higher score retain their position

5.5 Conclusions

Building knowledge networks to represent social aspect of communication across individuals within a community is a useful way to represent knowledge transfer and diffusion within the community. We can utilize the metrics of social network analysis to understand more about the underlying exchange of knowledge and individuals involved. We look at it in greater detail in Chapter 5.

Once we visualize such knowledge transfer, we also realize the natural cliques that form within a community. We look at this aspect of knowledge networks in greater detail in Chapter 6.

The analysis of the community for a particular period offers interesting insights. The important phrases within a community change over time. This result is significant because if expertise is sought in a particular area of interest, it would be preferable first to determine the period when it was considered an interesting phrase by the CoPs. Once this is known, it is easy to find experts in the relevant field or knowledge base from that era. If the total lifecycle of a CoP is examined, such topics may seem insignificant over the whole period.

Discussion of the individual within the network offers interesting insights. Three metrics for the individual are considered. Betweenness of individuals changes over the period, but once established as a matchmaker; the individual remains so until the time he or she is part of the group. In most cases, the relative ranking of the individual remains intact. It is evident that indegree changes over the period, but once an individual becomes established and sought after, he or she remains so. In most cases, the relative ranking remains intact until the individual becomes part of the group.

As with indegree, the outdegree changes over the period, but once an individual becomes established as a prolific contributor, he or she remains so. In most cases, the relative ranking remains intact until the individual becomes part of the group.

We looked at similar data from communications from multiple topics to make sure that it is not domain specific. Appendix E contains similar charts from sci.lang, sci.psychology, and rec.birds group. The conclusion is same, once an individual establishes himself, his remains there until the time he leaves the group.

Once the knowledge network is established, the next challenge is how to query that network to reach the individual with tacit knowledge that is of interest. The first challenge is identifying appropriate keywords based on what is of interest when performing a search. It is also necessary to understand that knowledge may be concentrated in few individuals. Chapter 6 examines the concept of knowledge adjacency and cliques so that the individual with tacit knowledge can be reached effectively. Moreover, individuals holding a community together and the impact of any individual leaving the community is of relative importance.

CHAPTER 6

6. Knowledge Adjacency and Cliques

6.1 Knowledge adjacency

A significant amount of communication that takes places across CoPs consists of specialized taxonomy and vocabulary. When dealing with such content, a mechanism needs to be built for finding phrases that are similar in meaning. The concept of knowledge adjacency was defined to work around this problem.

Adjacent phrases are simply phrases that are more likely to occur together. Phrases, occurring together within the body of a single message, would be assumed to be adjacent to each other. The higher the frequency of such phrases occurring together, the higher their adjacency and lower their distance.

6.1.1 The problem

The problem of knowledge discovery is complex since the individual may not know the appropriate phrase that will lead to what is being sought. Because of this, it is important to determine which phrases are related to each other. It is important to ensure that both the phrases in a pair exist in the same message so that their occurrences together can be computed. Once the phrase pairs and their count are known, these over all the messages are aggregated to help with the computation of the conditional probability of a second phrase occurring given the fact the first phrase in the pair has occurred. A phrase pair, with higher conditional probability, results in a lower distance between the two phrases of that pair. Phrases that have very low distance from a given phrase are within the knowledge adjacency of that phrase. Every message is processed with the above algorithm, and then it is possible to find in the top phrases in a topic the distance of other phrases from a phrase.

6.1.2 The solution

The very first step that needs to be taken is to find what are the top phrases in a body of knowledge. Table 11 gives an example of top 10 phrases extracted from top six topics from the dataset. Table 12 lists the distance of phrases from the given phrase in a topic. In topics like literature or general news, one can safely substitute words with similar meaning to search, but in specialized content, such as content related to technology, legal and medicine, etc. a significant amount of jargon is used. Synonyms are not good candidates for alternative keywords while searching for content. For example, C++ language uses the class as a keyword, the synonyms of class such as category, grade, rating, classification, etc. are not a good alternative to class when searching for content in C++.

Table 11 Top phrases in various topics

Rank	Comp.lang.c++	comp.lang.c	sci.phychology	alt.religion	sci.electronics	sci.physics
1	class	int	mail	new	do	new
2	int	do	inform	now	time	do
3	do	code	like	time	get	time
4	new	char	system	like	power	make
5	function	return	univers	someth	like	same
6	include	function	scienc	theori	system	messag
7	code	include	work	thing	work	physic
8	return	program	comput	differ	10	two
9	void	compil	research	believ	line	even
10	object	write	gener	set	signal	energi

Table 11 enumerates the list of the top 10 phrases in various topics.

A stop list of words was used to eliminate frequently occurring common words, but many of the words do have meaning in some of the contexts. For example, the word *do* is specific meaning in communities like *comp.lang.c* and *comp.lang.c++* while in communities like *sci.physics* it doesn't have any specific meaning. We also see word *10* listed in top words. On examination of underlying messages, we see that the occurrence of *10* in *sci.electronics* carries specific meaning. Following are some of the examples of sentences from messages which have specific meaning in the context of *sci.electronics*.

Pin 10 on the IF amplifier is an "S" meter output.
around 10 minutes of running video or 30*60*10 frames or twice
at 10 Mhz. No problem. However, using today's faster logics (F, AS,...)
Note that equivalent arrays would also be okay [six 10 ohm 100-200W, etc] as

Clearly the word *10* is being used in electronics message where it has meaning, if we compare this to occurrence of *do* in physics, we observe occurrence of following types of messages.

Analog components are nice because sometimes they do just what you
Why do FM radio's get better and then worse reception as you walk
What I'm wondering whether this exercise can do is replace the
mechanism of hearing you should do the same to the dynamic range

The above occurrences of word *do* is clearly being used in the context of English language and does not have very specific meaning.

```
do {
#define PUSH(LOW,HIGH) do {top->lo = LOW;top->hi = HIGH;top++;} while (0)
#define POP(LOW,HIGH) do {--top;LOW = top->lo;HIGH = top->hi;} while (0)
#define SWAP(A,B) do {int_temp = (A);(A) = (B);(B) =_temp;} while (0)
of what you can do in the future. Technology is constantly changing. If
I didn't understand the comment about "less special casing;" they do require
```

Similarly, we look at occurrence of do in community *comp.lang.c++* and we see a combination of the word being used for its English meaning and in the context of a specific meaning as defined in the C++ language.

The occurrence of word *two* in context of community *sci.physics* also reveals the usage of word in a context where it has special meaning.

way through, the time rate difference between the two halves would, I
(1) Finding the distance betwen two points in a graph can be solved
with the walls in two layers. The inner layer being thermally conductive
There are two principal designs of REGs in use as well. One, the Schmidt
phase velocities of the two broadcast waves. Let's add the waves again:

Table 12 Phrase distance from top phrase

Rank	Comp.lang.c++	comp.lang.c	sci.phychology	alt.religion	sci.electronics	sci.physics
From Phrase	class	int	mail	do	new	energi
1	int	return	psychnet	time	get	time
2	function	type	send	physic	time	mass
3	public	pointer	request	thing	like	new
4	object	value	node	world	make	physic
5	void	void	univers	realiti	work	space
6	virtual	printf	research	postmodern	power	particl
7	type	program	psycholog	plain	system	field
8	const	null	system	pint	signal	photon
9	oper	struct	program	people	me	univers

Table 12 takes one phrase as an example and lists the phrases that have the shortest distance from that particular phrase. The result denotes the probability of the phrases occurring together. For example in *sci.physics*, when the discussion is about *energi*, the *time* occurs with that phrase most often.

The problem is common English words is also eliminated once we use a word along with its adjacent word. As we can see in Table 12, if we choose a phrase that is relevant to that domain for sure, the common English words are not close to that word.

We propose that words that fall within the knowledge adjacency of a given phrase are more appropriate alternatives for a keyword when searching for knowledge. We need to search the keyword and other phrases that fall within the knowledge adjacency of that phrase. Therefore, if one is searching for *energi*, the phrase *time* may be the second-best proxy for that phrase.

Figure 29 shows the word cloud of topic *sci.physics* during July 2003. The phrases in the word cloud are stemmed phrases.

Figure 29 Word cloud for sci.physics for a particular period; the size of the word shows the relative importance of the word during the period

Figure 30 provides a way to visualize the knowledge adjacency. The word for which adjacent words are sought is in the center, and the rest of the phrases are laid out based on their distance.

The size of the bubble is also based on the distance. Therefore, larger the bubble, the closer the word is to the original word. We chose a word from the word cloud *energi* and generate the knowledge adjacency of this phrase. As we can see the closest phrases are *field, see, human, frequency, radiation*. In the context of physics, if we are searching for *energi*, it may be beneficial also to search for above phrases in its knowledge adjacency to obtain a more accurate result.

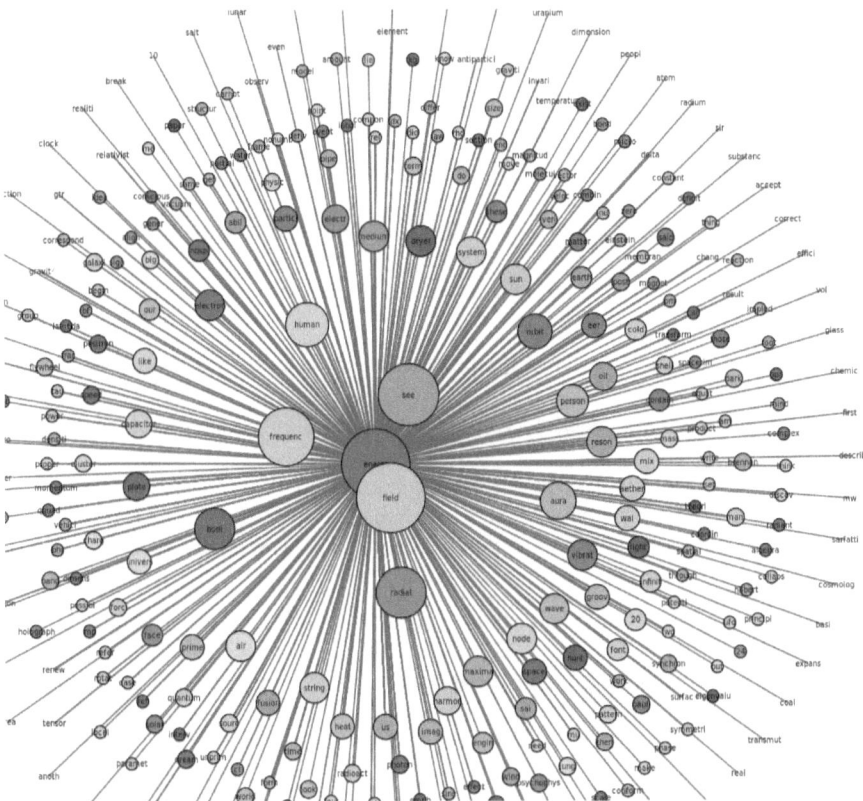

Figure 30 Adjacency of phrases; the distance and the size of bubble shows the distance of the word from the word in center

6.2 Cliques

In this section, we examine at the natural cliques that form in large communities. We define cohesiveness of the clique as the fraction of communication that is from group members. If there existed a single clique within a community whose members only communicated amongst themselves, then their cohesiveness would be 1.

Cohesiveness of clique $C_{cq} = \dfrac{T_g \text{ (Total communication by clique members)}}{T \text{ (Total communication by all individuals in community)}}$

$$= \frac{\sum Outdegree_{clique}}{\sum Outdegree_{total}}.$$

Cohesiveness of communication $C_c = \dfrac{T_t \text{ (Total communication related to a phrase)}}{T_g \text{ (Total communication about all phrases)}}$

$$= \frac{\sum EdgeWeights_{phrase}}{\sum EdgeWeights_{total}}$$

The cohesiveness of the communication can be defined as the fraction of communication that is related to a phrase. If there exists a group that only communicates about a single phrase, then the cohesiveness of communication about that phrase within that group would be 1.

The reason that we have classified the communication based on cohesiveness because it has special meaning. There are two specific types of communications that we are extremely interested in. In Figure 5, right top and right bottom quadrants are of interest to us. Following are the problems of interest that we intend to explore in the above two areas.

1. Highly cohesive communication among highly cohesive cliques points to the knowledge being exchanged by the experts and the interested parties of that particular topic. It is expected that the knowledge exchange in such cases is of very high quality and is extremely beneficial for learning the point of view. Most of the corporations are interested in such communication where a very high cohesive group is exchanging information about a topic. Furthermore, this is also the best source to find tacit knowledge related to that topic and adjacencies.

2. Highly cohesive cliques communicating with very low cohesiveness communication points to special interest groups. These are of extreme interest to find out

communications related to the topic that may not be of interest to a very large number of people but are very important. Such conversation is particularly useful to groups like security agencies. For example, few individuals communicating in code language among each other may not mean anything to anybody else but could trigger interest from security agencies to monitor that discussion again.

Figure 31 depicts a knowledge network created from communication in sci.physics during the half-year period starting July 1986. As can be noted, there is a large dense cluster of nodes to which most of the nodes are connected with some nodes forming their clusters and completely disconnected from the rest of the nodes.

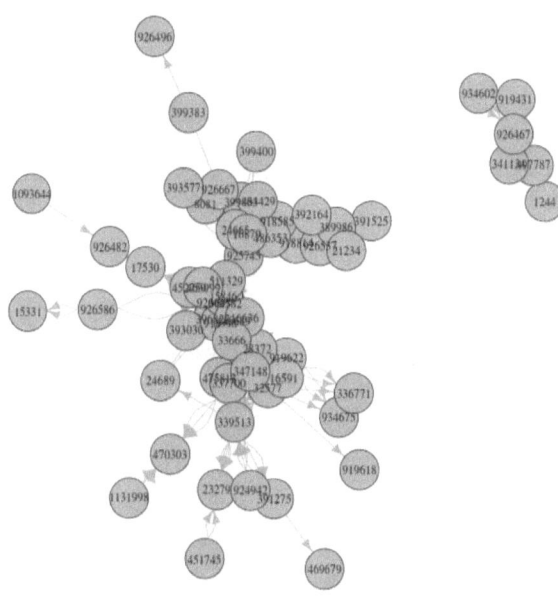

Figure 31 Sci.physics network for the half year starting July 1986; the obvious cliques are visible

6.2.1 Which algorithm?

Multiple algorithms in the literature can be used in the detection of cliques. We experiment with the following algorithms to decide which one of them works best in our scenario.

6.2.1.1 *Infomap community detection*

We examined multiple algorithms for clique detection within a knowledge network. The first algorithm that we used is Infomap community detection which uses the probability flow of random walks on a network as a proxy for information flows in the real system and decompose the network into modules by compressing a description of the probability flow(Rosvall & Bergstrom, 2008)

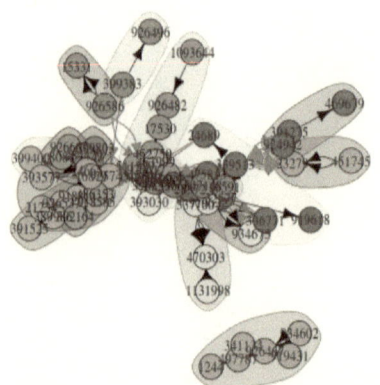

Figure 32 Infomap community detection algorithm

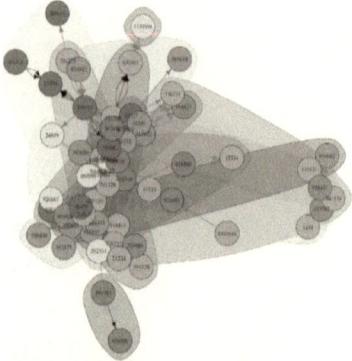

Figure 33 Infomap Community detection after deleting top betweenness element

Figure 32 depicts the output of Infomap community detection algorithm on the network.

Since the objective was to evaluate the impact of the top member leaving the community, we delete the node with the highest betweenness and re-run the Infomap community detection algorithm. Figure 33 depicts the output of the algorithm.

6.2.1.2 Fastgreedy community detection

Fast greedy community detection could not be used because it does not support multiple edges across two nodes. That is the basic structure of our knowledge graph(Clauset, Newman, & Moore, 2004).

6.2.1.3 Blondel's multilevel community detection algorithm

Figure 34 depicts the output of community detection using Blondel's algorithm. The value to be optimized is modularity, defined as a value between -1 and 1 that measures the density of links inside communities compared to links between communities. First, each node in the network is assigned to its own community. Then for each node i, the change in modularity is calculated for removing i from its own community and moving it into the community of each neighbor j of i.

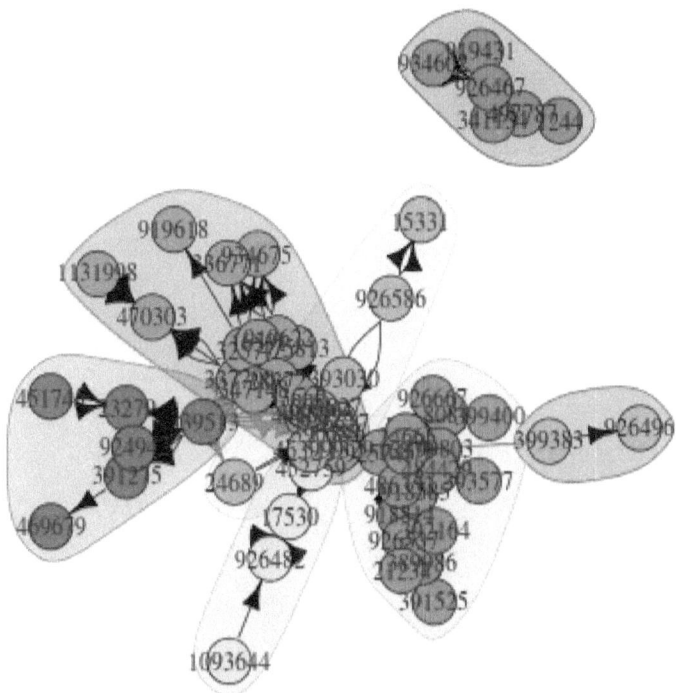

Figure 34 sci.physics community cliques using Blondel's algorithm; additional sub-networks are found

In order to perform clique detection analysis on a graph, we need to take following steps.

- Generate the graph

- Detect cliques within the graph

For example, we take a graph from the sci.physics group for the period July 1986 to January 1987. Figure 31 has the visual representation of the graph. As we can see some of the cliques are obvious to locate, while others are not so obvious.

Next, we run the multilevel community detection algorithm (Blondel et al., 2008) on the graph and get a graph where communities are annotated. As we can see the obvious grouping of nodes not connected to the rest of graph forms a clique, but there are other nodes which were not so obvious from the basic graph result into their cliques. Based on performance and effectiveness, for further analysis, we only use Blondel's algorithm. Here are all the graphs of the communities in the above graph.

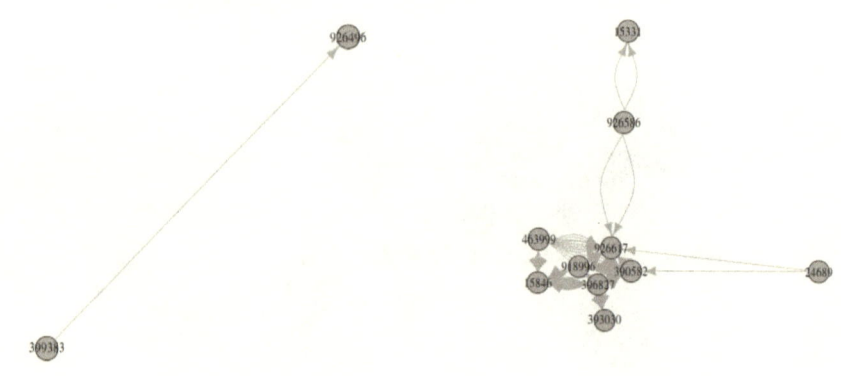

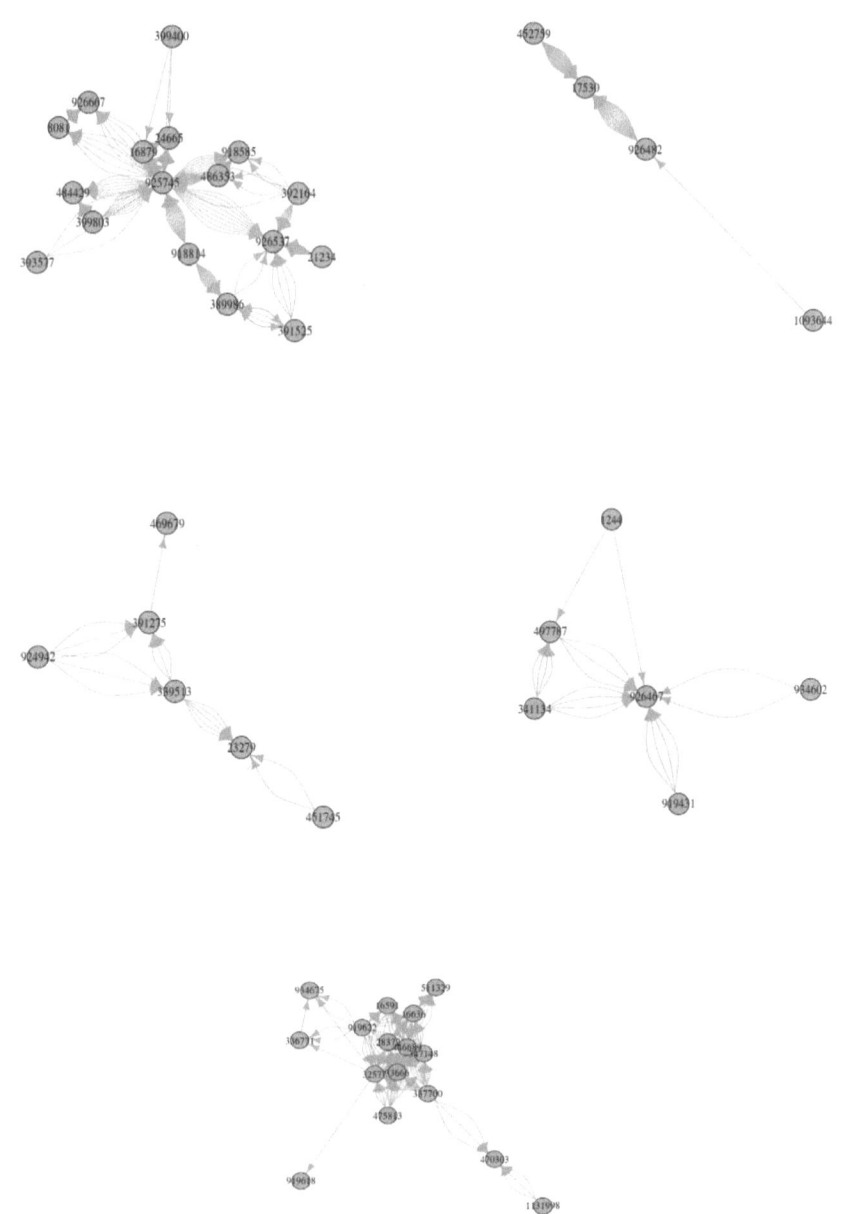

Figure 35 Cliques within the community as identified by the algorithm on given network

Figure 35 enumerates seven cliques detected by the Blondel's algorithm. As we have discussed

earlier in this section, natural cliques form within a topic based on the interaction between individuals. These cliques consist of individuals who have the higher cohesiveness of communication within themselves compared to individuals who are in different cliques. The prominent cliques can define a characteristic of a topic within a topic. Since knowledge graphs are directed graphs, we calculate at the outdegree of each of the user within the cliques and calculate cohesiveness of different cliques within a topic.

6.2.2 Cohesiveness of a clique

Figure 36 depicts natural cliques found within sci.physics community during the half year starting from July 1986. Group1, Group4, Group5, and Group6 can be identified as cliques which contribute least to the topic.

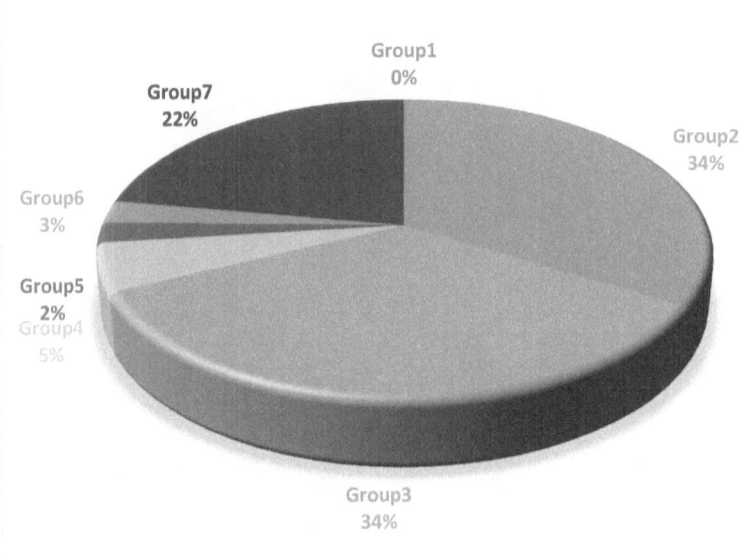

Figure 36 Contribution of cliques to community; Group1 to Group 7 contribute 0%, 34%, 34%, 5%, 2%, 3%, and 22% respectively to the community

The sci.physics community for a particular period was analyzed as described below and this resulted in seven distinct cliques as shown above. As illustrated in the scatter chart in Figure

37, most of the phrases belong to a single clique; while some belong to two. Not many phrases are spread over more than two cliques.

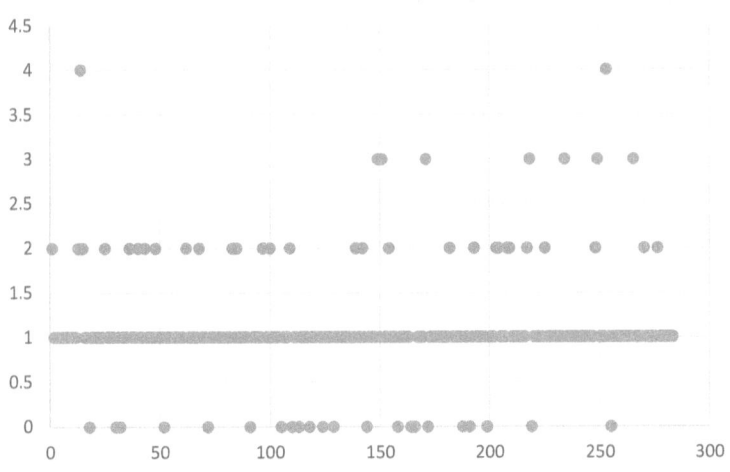

Figure 37 Phrase representation in cliques; majority of phrases belong to just one clique, very few belong to more than two cliques

The detailed results related to phrase representation within the community are given in Appendix D.

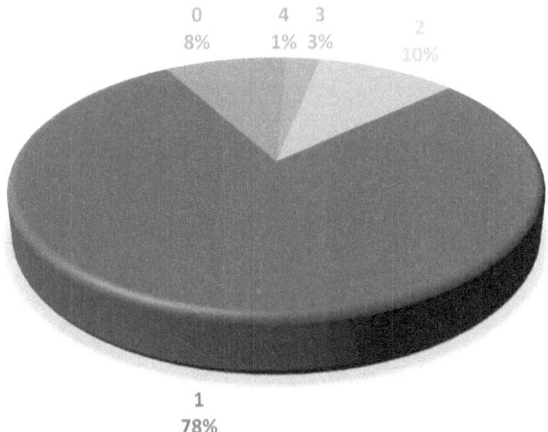

Figure 38 Phrase representation distribution; As much as 78% phrases belong to one clique

Figure 37 and Figure 38 depict that most of the phrases belong to a single clique within a community. Figure 39 plots the cohesiveness of communication and groups on a per phrase basis. The phrase and group combination that has zero contribution are eliminated. For clarity sake, the data table is produced below in Table 13.

Table 13 Cohesiveness of groups and communication

phrase	Cohesiveness of clique	Cohesiveness of communication
acceler(group2)	33.46%	26.89%
acceler(group7)	22.22%	7.14%
articl(group3)	34.36%	7.89%
articl(group7)	22.22%	92.11%
defeat(group7)	22.22%	100.00%
filler(group7)	22.22%	100.00%
suppress(group7)	22.22%	100.00%
know(group3)	34.36%	62.21%
rider(group2)	33.46%	54.74%
time(group2)	33.46%	41.88%
beta(group7)	22.22%	37.11%

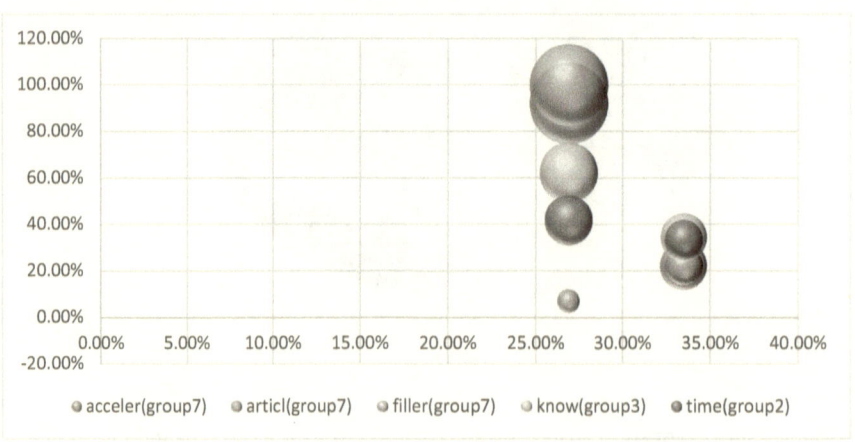

Figure 39 Cohesiveness of communications and cliques; top phrases belong in third (interesting phrases) or fourth (phrases of interest) quadrant

The Y-axis contains the cohesiveness of communication, while the X axis contains the cohesiveness of a clique. We have chosen top 9 phrases and plotted the graph for those. As we can see, referring to Figure 5, top phrases belong either in third (interesting topics) or fourth (topics of interest) quadrant. The cohesiveness of any clique here is not very high, and because of that a low cohesiveness of communication just means that the phrase is just too common and many people are interested in it.

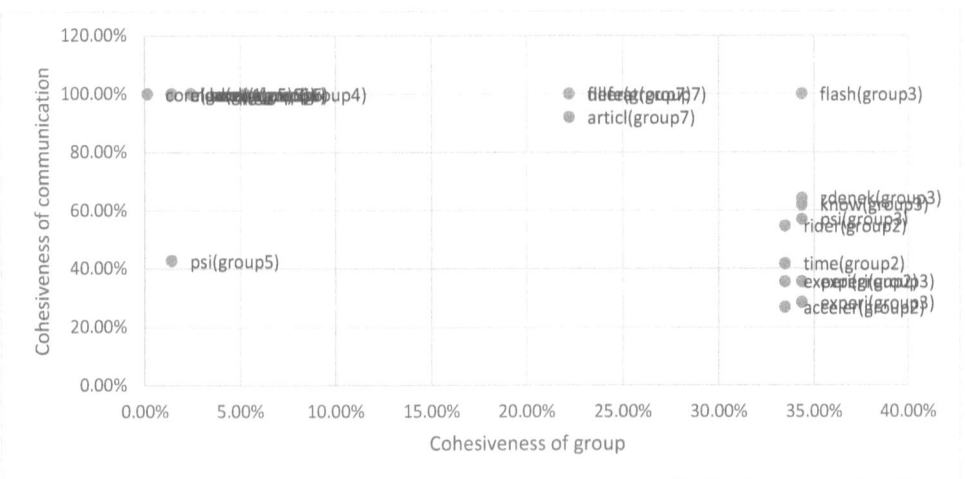

Figure 40 Top phrases in clique, Cohesiveness; top phrases belong in third and fourth quadrant

In Figure 40 We take another sample, in this scenario, rather than choosing top phrases, we choose top phrases from each group. Now we can see that we find more cliques are represented in the graph. The conclusions are similar.

6.2.3   Community Attrition

Communities that have fewer members with very high centrality run the risk of disintegrating when that key member leaves. In a typical community when such key individuals leave, they also carry a significant amount of tacit knowledge with themselves leaving the community with a knowledge deficit that is extremely difficult to bridge.

Figure 41 consists of a community that consists of 12 naturally occurring cliques. This community consists of 77 members represented by nodes with 132 edges connecting them with each other.

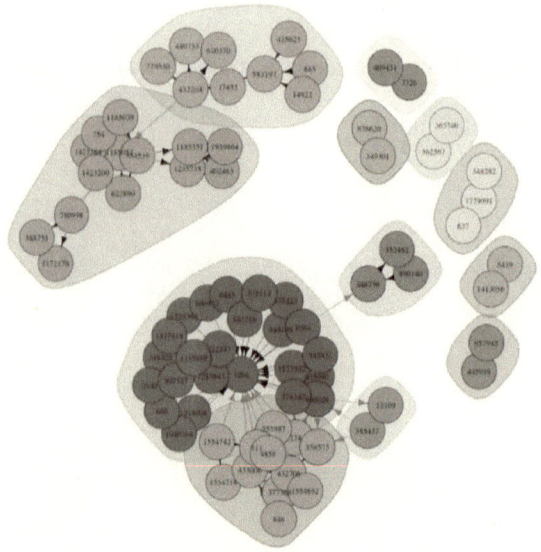

Figure 41 A community with 12 cliques

We find out the node with the highest centrality within the community and delete it from the graph to simulate the situation when the member with highest centrality leaves.

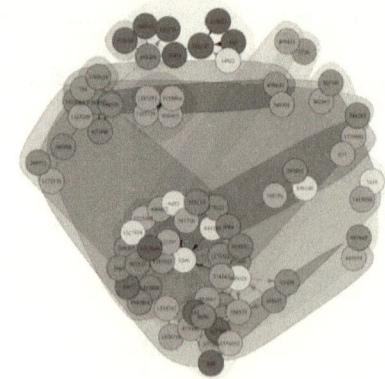

Figure 42 Community when top member leaves; number of cliques increase to 22

Figure 42 describes the situation when the top member leaves. As we can see, just the attrition of top member results in the number of natural groups to increase to 22. One top member was holding almost ten groups together.

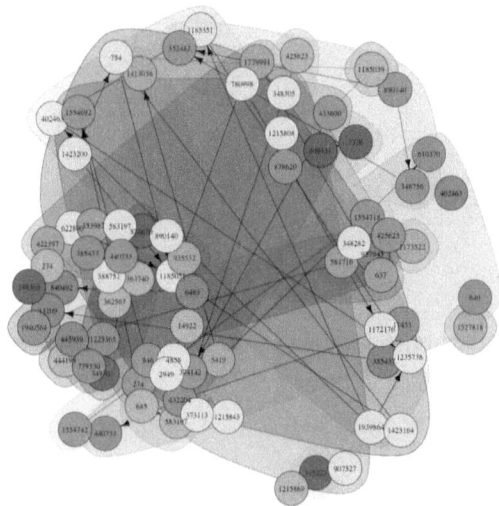

Figure 43 Community when ten members leave; number of cliques become 28

Similarly, if we look at Figure 43, when additional ten members leave, the community now consists of 28 naturally occurring cliques. The community stands further splintered.

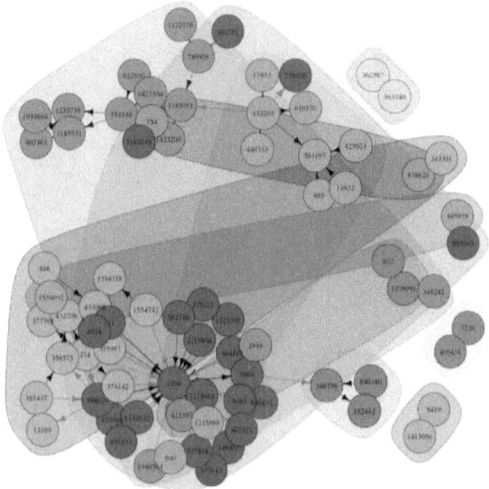

Figure 44 Community when individual with the highest outdegree leaves; number of cliques increase to 16

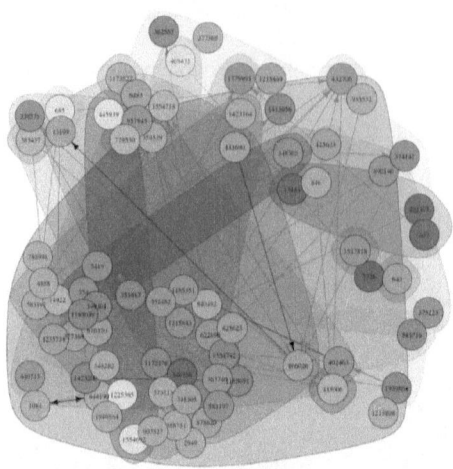

Figure 45 Community when top 5 members with the highest outdegree leave; number of cliques increase to 26

We see a similar splintering of the network when individuals with high outdegree leave but the impact is not as significant as when an individual with high betweenness leaves. As we can see from Figure 44, when the individual with highest outdegree leaves, the graph improves its cohesiveness because the number of natural cliques reduces to 11 from 12 but when as shown in Figure 45 five members leave, the number of cliques in the community increases to 26.

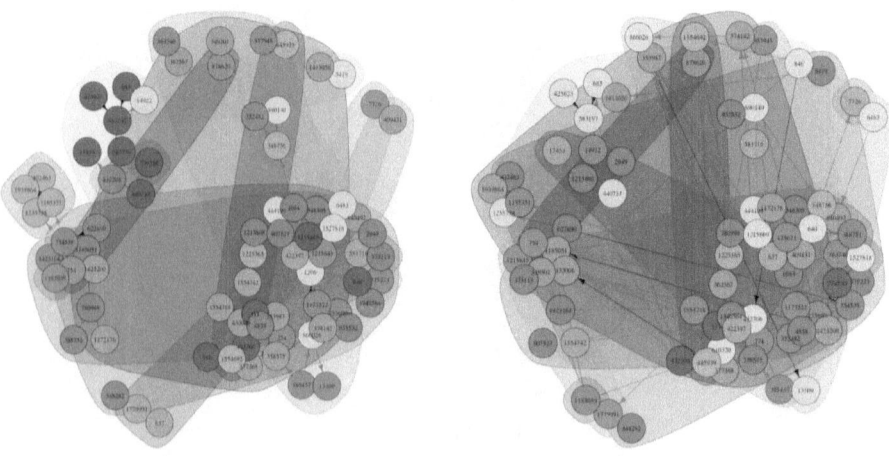

Figure 46 Community when member with the highest indegree leaves

Figure 47 Community when top two members with the highest indegree leave

We did a similar exercise with members with top indegree, and we see in Figure 46 Community when member with the highest indegree leaves, when the first member with top indegree leaves, the graph in intact but as shown in Figure 47 Community when top two members with the highest indegree leave when second member leaves, the groups within community increases from 12 to 22.

6.3 Conclusions

We establish that using similar meaning words in the context of the body of knowledge that is full of specialized taxonomy and vocabulary consisting of technical terms may not be the best approach while searching for a keyword in a body of knowledge. We propose that words with higher knowledge adjacency index compared to a phrase of interest may offer a better choice of alternative phrases when one wants to search a given body of knowledge. Knowledge adjacency index should also be used keeping particular time-periods in mind because adjacent words also change over a period.

Once we have a knowledge network constructed from a body of knowledge, finding inherent cliques is a very important tool to determine the nature of discussions that are happening within that community. Also, each of these naturally forming cliques have their representative phrases. As much as 78% phrases are seen in one clique at a time with as much as 90% accounting for not more than two cliques.

The representative phrases for these cliques can be used to classify them cliques into cliques of expertise, cliques of niche discussions or cliques of irrelevant discussions. For example, if a clique of individuals is discussing movies while the community is dedicated to the discussion of physics, the representative phrases would point to that fact.

Communications containing phrases with the very high cohesiveness of cliques and high cohesiveness of communication point to a discussion among experts. Once we have found the group with expertise with knowledge represented by a phrase, we can use social network analysis techniques to find the individuals with high betweenness, outdegree, and indegree to reach a real expert.

We also evaluate the impact of members with higher betweenness, outdegree and indegree leave the community. We conclude that if any of the members with high betweenness, outdegree and indegree leaves the community it has a significant impact on graph leading it to become less cohesive.

Chapter 4, Chapter 5, and Chapter 6 describe a complete methodology of building knowledge networks, the concept of knowledge adjacency to help us in choosing right keywords, the concept of cliques to help us in identifying appropriate individuals based on keywords that we are searching. In Chapter 7, we summarize the complete methodology and show how we have provided solutions to research questions that we set out to examine. Appendix A defines the architecture of a system that we had built to implement the concepts, methodologies, and algorithms that we have discussed in the above chapters.

CHAPTER 7

7. Conclusions, Limitations, and Future Research

7.1 Conclusions

This thesis presents the research work done to evaluate methodology to extract knowledge and source of knowledge from a communication repository. The premise is that most of the knowledge is residing in the communication archives of organizations and CoPs. The approach to reach the tacit knowledge that is in individuals involved is to locate those individuals.

In Chapter 4, the objective was to validate the existing belief that organizations lose some of the contained tacit knowledge when individuals leave an organization. A set of hypotheses were established and were validated through a survey and subsequent quantitative analysis. Following conclusions were drawn at the end of this activity.

1. Subject matter experts would seek help from others which includes individuals within the group, individuals outside the group, individuals within the organization, and individuals outside the organization.

2. The knowledge of experts diffuses into their closest team members. They are as likely and capable of providing help when needed.

3. Individuals who are asked for help are also more likely to ask for help pointing to a cultural aspect of teams. Individuals who see the value of helping others both seek and provide help to others.

4. Individuals who have experts in their team are more likely to ask experts in other teams as well pointing to the fact that the value of expert advice is well understood in teams which work with experts on a regular basis.

5. We also concluded that Organization and process experts do not seek help from others.

In Chapter 5 we presented a framework that was required to work on next research problem. We defined the methodology and algorithm to build knowledge networks from communication repository. The knowledge networks which are the output of this activity represent the underlying relationship across individuals due to interactions between them. It also captures the social aspect of knowledge transfer. Social Network Analysis tools can be easily applied to such knowledge network, and they can provide us valuable insights into how the knowledge is organized and how it is being exchanged between the individuals. Since we have used Nonaka's model of tacit knowledge, social aspect also explains in the transfer of some of the tacit knowledge through the process of socialization(Ikujiro Nonaka, 2007). We set out to evaluate the evolution of knowledge within the communities, groups, and individuals from the knowledge networks that were built using above methodology. We have experimented with knowledge graphs, our test dataset (Spencer & Wiseman, n.d.) and the tools of social network analysis and draw inferences on how the communities and individuals evolve over a period. Our conclusions are as follows.

1. Representative phrases are identifiers of communities. The phrases themselves Do not remain constant over a long period but provide good insight into what a community is in a particular period.

2. Representative phrases of a community form a continuum for that community and are significant to search for relevant information within that community.

3. Individuals, like communities, also evolve over a period. We classified individuals based on three primary measures, betweenness, outdegree and indegree into matchmakers, experts and knowledge seekers respectively. It was observed that individuals who attain higher positions based on any of those measures remain at higher positions till the time they leave the group.

In Chapter 6 we set out to build on the concept of knowledge adjacency. We defined a methodology and algorithm to generate knowledge adjacency. It was established that for knowledge that consists of special taxonomy and vocabulary and domain-specific phrases, language-based synonyms might not be best alternative phrases when one is looking to search for something. A methodology and algorithm were presented for the computation of knowledge adjacency. We also defined methods for visualization to quickly traverse through knowledge through phrases and phrases in knowledge adjacency of that phrase. Our conclusions from this work are as follows.

1. Phrases which form edges of knowledge graph, coupled with phrases in the knowledge adjacency of the phrase of interest would provide pointers to most appropriate knowledge as well as the individual who has expertise in it.

2. Knowledge graphs also define a working relationship among individuals within a community. For example, an individual may have great engagement with another individual in the field of physics while he may have a very different engagement with another individual in the field of automobiles because he pursues it as a hobby.

3. Knowledge graphs also provide a way for an individual to get in touch with an expert even though he may not be directly interacting with him.

We set out to explore the phenomenon of naturally forming cliques within communities(Simon, 1950). We looked at the representative phrases for cliques. We established that most cliques have a pretty unique set of representative phrases and a look at these phrases would tell us how a particular clique is distinct from other cliques. A methodology was also presented to compute cohesiveness of cliques and communication to help in classification of cliques. Our conclusions from this activity are as follows.

1. Naturally occurring inherent cliques are a very important tool to find the nature of discussions that are happening within that community.

2. Each of these naturally forming cliques have their representative phrases. As large as 78% phrases are only seen in one clique at a time with as much as 90% accounting for not more than two cliques.

3. Representative phrases for these cliques can be used to classify these cliques into cliques of expertise, cliques of niche discussions or cliques of irrelevant discussions.

4. Discussions containing phrases that have the high cohesiveness of clique and high cohesiveness of communication point to a discussion among experts.

5. Once we have found the clique with expertise with knowledge represented by a phrase, we can use social network analysis techniques to find the individuals with high centrality.

We also look at what happens when individuals in the leadership position within the community leave. When individuals with high betweenness, outdegree, and in degree leave the community, the community starts to get splintered. We observe more naturally occurring cliques being created. Another significant result is that it does not matter whether the individual had high betweenness, outdegree, or indegree, the results are similar. The observation leads us to a conclusion that whether the individual is a high contributor, high matchmaker or the individual who is asking most questions, all of them are equally important to keep the structure of community intact. If any of them leave leadership position, it is harmful to the community.

In Appendix A all the methodology and algorithms that were developed as part of this research work were consolidated into an extensible system architecture that can be used to build a system that does all the analysis on any given communication repository. We have enumerated an architecture that has been evolved from multiple pieces of algorithms that we had to develop to complete this research. The architecture can be extended to add more types of communication repositories.

The results obtained through this research are applicable across multiple domains because different newsgroups in Usenet have very different content. Appendix E articulates results from an additional newsgroup which is very dissimilar, and we find that these are very like each other.

7.2 Benefits

The system and architecture proposed as part of this research along with the set of algorithms developed to provide an elegant way to search the source of knowledge in a community of practice. Many knowledge management systems provide the functionality for indexing content and searching through the content with keywords. Our research is different and significant in following ways.

1. We provide a mechanism to identify the period when any topic of interest was most popular. What a user is looking to search may have been most popular many years ago, and any of the recent content may not provide sufficient insights into it. For example, the field was AI saw lots of investment and research from 1956 until the early 1970s, then suddenly around 1974 the progress slowed significantly. The field was revived in the early 1980s. So, if a person were to look for AI-related research during the late 1970s, he would find very little evidence. Our system can tell the user that the best experts are from the era of the early 1970s.

2. While searching through content, synonyms are not very appropriate in technical content because they consist of special taxonomy and vocabulary. We propose a mechanism of knowledge adjacency that provides a good mechanism to find what needs to be searched depending on the domain of the search.

3. Our system helps in identification of key individuals who are holding a community together. The departure of these individuals can significantly damage a community.

4. Our system helps in identifying cliques in the community thus helping in the identification of side conversations that are happening in a community.

7.3 Limitations

We have not looked at any audio and video processing algorithms that can be used to process such content. We have also not looked at any video to text and audio to text converters that can be used to convert such information into text and processed. If such converter were to be available, the architecture given in Appendix A has the flexibility to incorporate it and process such communication. We have also not looked at the problems associated with scaling by utilizing any of the cloud technologies or distributed computing technologies.

We have also not done any deep language analysis to compute the quality of communication that is exchanged between individuals.

7.4 Future Research

Most of the limitations presented in the previous section can be taken up as future research opportunities. Apart from these opportunities the problem space and current work lends itself very well to enhance it to build an automated credibility system that can be used to rank individuals participating in such communication. Additional areas of research include identifying misclassified messages by textual analysis and classifying them correctly, better identification of common phrases that need to be eliminated based on underlying topic, the capability to handle different languages.

REFERENCES

Akbar, H. (2003). Knowledge Levels and their Transformation: Towards the Integration of Knowledge Creation and Individual Learning. *Journal of Management Studies, 40*(8), 1997–2021. https://doi.org/10.1046/j.1467-6486.2003.00409.x

Alavi, M., & Leidner, D. E. (2001). Knowledge Management and Knowledge Management Systems: Conceptual Foundations and Research Issues. *MIS Quarterly, 25*(1), 107–136.

Alias-i. (2008). LingPipe 4.1.0. Retrieved from http://alias-i.com/lingpipe

Alipour, F., Idris, K., & Karimi, R. (2011). Knowledge Creation and Transfer: Role of Learning Organization. *International Journal of Business Administration, 2*(3), 61–67. https://doi.org/10.5430/ijba.v2n3p61

Anand, V., Manz, C. C., Glick, W. H., & Glick, H. (1998). NOTE APPROACH MEMORY MANAGEMENT TO INFORMATION University of Massachusetts. *The Academy of Management Review, 23*(4), 796–809.

Ardichvili, A., Page, V., & Wentling, T. (2003). Motivation and barriers to participation in virtual knowledge-sharing communities of practice. *Journal of Knowledge Management, 7*(1), 64–77. https://doi.org/10.1108/13673270310463626

Argote, L. (2005). Reflections on Two Views of Managing Learning and Knowledge in Organizations. *Journal of Management Inquiry, 14*(1), 43–48. https://doi.org/10.1177/1056492604273179

Argyris, C. (1977). Double loop learning in organizations. *Harvard Business Review, 55*(5), 115–125. https://doi.org/10.1007/BF02013415

Avasthi, Vinay, & Dey, S. (2015). Loss in tacit knowledge because of employees attrition Vinay

Avasthi Shubhamoy Dey. *Internation Journal of Intercultural Information Management*, *5*(x), 1–18. https://doi.org/10.1504/IJIIM.2015.072541

Avasthi, V., Dey, S., Jain, K. K., & Mishra, R. (2015). The Evolution of Knowledge in Communities of Practice. *Proceedings of the 2015 Conference on Research in Adaptive and Convergent Systems*, 1–6. https://doi.org/10.1145/2811411.2811528

Bakker, M., Leenders, R. T. A. J., Gabbay, S. M., Kratzer, J., & Engelen, J. M. L. Van. (2006). Is trust really social capital? Knowledge sharing in product development projects. *The Learning Organization*, *13*(6), 594–605. https://doi.org/10.1108/09696470610705479

Becerra, M., Lunnan, R., & Huemer, L. (2008). Trustworthiness, risk, and the transfer of tacit and explicit knowledge between alliance partners. *Journal of Management Studies*, *45*(4), 691–713. https://doi.org/10.1111/j.1467-6486.2008.00766.x

Bishop, C. (2006). *Pattern Recognition and Machine Learning*. (M. Jordan, J. Kleinberg, & B. Schölkopf, Eds.), *Zhurnal Eksperimental'noi i Teoreticheskoi Fiziki* (Vol. 4). Springer. https://doi.org/10.1117/1.2819119

Blondel, V. D., Guillaume, J.-L., Lambiotte, R., & Lefebvre, E. (2008). Fast unfolding of communities in large networks. *Journal of Statistical Mechanics: Theory and Experiment*, *2008*(10), 2008. https://doi.org/10.1088/1742-5468/2008/10/P10008

Boland, R. J., Singh, J., Salipante, P., Aram, J. D., Fay, S. Y., & Kanawattanachai, P. (2001). Knowledge representations and knowlege transfer. *Academy of Management Journal*, *44*(2), 393–417. https://doi.org/citeulike-article-id:2934942

Bostock, M. (2012). D3. js. *Data Driven Documents*.

Bosua, R. (2013). Aligning strategies and processes in knowledge management: a framework. *Journal of Knowledge Management*, *17*(3), 331–346. https://doi.org/10.1108/JKM-10-

Brown, J. S., & Duguid, P. (1991). Toward a Unified View of Working, Learning, and Innovation. *Organization Science, 2*(1), 40–57. https://doi.org/10.1287/orsc.2.1.40

Burkhard, R. J., Hill, T. R., & Venkatsubramanyan, S. (2011). The Emerging Challenge of Knowledge Management Ecosystems: A Silicon Valley High Tech Company Signals the Future. *Information Systems Management, 28*(1), 5–18. https://doi.org/10.1080/10580530.2011.536105

Burlton, R. T. (2001). *Business Process Management: Profiting from process. Pennsylvania: Sams. 2001.*

Bush, A. A. (2005). Designing Sticky Knowledge Networks. *Communications of the ACM, 48*(5), 66–71. https://doi.org/10.1145/1060710.1060711

Carlile, P. (2002). A pragmatic view of knowledge and boundaries: Boundary objects in new product development. *Organization Science, 13*(4), 442–455. Retrieved from http://pubsonline.informs.org/doi/abs/10.1287/orsc.13.4.442.2953

Chatzkel, J. (2004). Establishing a global KM initiative: the Wipro story. *Journal of Knowledge Management, 8*(2), 6–18. https://doi.org/10.1108/13673270410529073

Chau, T., Maurer, F., & Melnik, G. (2003). Knowledge Sharing : Agile Methods vs . Tayloristic Methods 4 . Knowledge Sharing Support in Agile and.

Chua, A., & Lam, W. (2005). Why KM projects fail: a multi-case analysis. *Journal of Knowledge Management, 9*(3), 6–17. https://doi.org/10.1108/13673270510602737

Clauset, A., Newman, M. E. J., & Moore, C. (2004). Finding community structure in very large networks. *Physical Review E - Statistical, Nonlinear, and Soft Matter Physics, 70*(6 2), 66111. https://doi.org/10.1103/PhysRevE.70.066111

Coff, R. W., Coff, D. C., & Eastvold, R. (2006). The knowledge-leveraging paradox: How to achieve scale without making knowledge imitable. *Academy of Management Review*, *31*(2), 452–465. https://doi.org/10.5465/AMR.2006.20208690

Connell, N., Klein, J., & Powell, P. (2003). It's tacit knowledge but not as we know it: redirecting the search for knowledge. *Journal of the Operational Research Society*, *54*(2), 140–152. https://doi.org/10.1057/palgrave.jors.2601444

Connelly, C. E., & Kelloway, K. E. (2003). Predictors of employees' perceptions of knowledge sharing cultures. *Leadership & Organization Development Journal*, *24*(5), 294–301. https://doi.org/10.1108/01437730310485815

Cook, S., & Brown, J. (1999). Bridging epistemologies: The generative dance between organizational knowledge and organizational knowing. *Organization Science*, *10*(4), 381–400. https://doi.org/10.1287/orsc.10.4.381

Cook, S. D. N., & Brown, J. S. (1999). Bridging Epistemologies: The Generative Dance Between Organizational Knowledge and Organizational Knowing. *Organization Science*, *10*(4), 381–400.

Crossan, M. M., Lane, H. W., & White, R. E. (1999). An Organizational Learning Freamwork: From Intuition to Institution. *Academy of Management Review*, *24*(3), 522–537. https://doi.org/10.5465/AMR.1999.2202135

Daft, R. L., & Weick, K. E. (1984). Toward a Model of Organizations as Interpretation Systems. *Academy of Management Review*, *9*(2), 284–295. https://doi.org/10.5465/AMR.1984.4277657

Davenport, T. H. (2005). *Thinking for a Living: how to get better performance and results from knowledge workers. Harvard Business School Press* (Vol. 18). Harvard Business School Press. https://doi.org/10.1002/hrdq.1221

Davenport, T. H., & Prusak, L. (1998). *Working Knowledge: How Organizations Manage What They Know. Knowledge Creation Diffusion Utilization*. Harvard Business Press. https://doi.org/10.1109/EMR.2003.1267012

Davenport, T. H., & Prusak, L. (2000). *Working Knowledge. How organizations manage what they know*. Harvard Business School Press. https://doi.org/10.1.1.11.6086

de Vries, R. E., van den Hooff, B., & de Ridder, J. a. (2006). Explaining Knowledge Sharing: The Role of Team Communication Styles, Job Satisfaction, and Performance Beliefs. *Communication Research*, *33*(2), 115–135. https://doi.org/10.1177/0093650205285366

Ellis, B., Stylos, J., & Myers, B. (2007). The factory pattern in API design: A usability evaluation. In *Proceedings - International Conference on Software Engineering* (pp. 302–311). https://doi.org/10.1109/ICSE.2007.85

Emelo, R. (2012). Why personal reputation matters in virtual knowledge sharing. *Industrial and Commercial Training*, *44*(1), 35–40. https://doi.org/10.1108/00197851211193408

Fang, Y., Jiang, G. L. F., Makino, S., & Beamish, P. W. (2010). Multinational firm knowledge, use of expatriates, and foreign subsidiary performance. *Journal of Management Studies*, *47*(1), 27–54. https://doi.org/10.1111/j.1467-6486.2009.00850.x

Fayyad, U., Piatetsky-Shapiro, G., & Smyth, P. (1996). Knowledge Discovery and Data Mining: Towards a Unifying Framework. In *Int Conf on Knowledge Discovery and Data Mining* (Vol. 96, pp. 82–88). https://doi.org/10.1.1.27.363

Ferber, R. (1977). Research By Convenience. *Journal of Consumer Research*, *4*(1), 57–58. https://doi.org/10.1086/208679

Foster, I., Kesselman, C., & Tuecke, S. (2001). The Anatomy of the Grid. *Hand Clinics*, *17*(4), 525–532. https://doi.org/10.1002/ar.1090120102

Gable, G. G. (2005). The enterprise system lifecycle: Through a knowledge management lens. *Strategic Change, 14*(August), 255–263. https://doi.org/http://dx.doi.org/10.1002/jsc.726

Gabor, C., & Tamas, N. (2006). The igraph software package for complex network research. *InterJournal, Complex Sy*, 1695. https://doi.org/citeulike-article-id:3443126

Girvan, M., & Newman, M. E. K. (2002). Community structure in social and biological networks. *Proct Natl Acad Sci USA, 99*(12), 7821–7826.

Grant, R. M. (1996a). Prospering in Dynamically-competitive Environments: Organizational Capability as Knowledge Integration. *Organization Science, 7*(4), 13. https://doi.org/10.1287/orsc.7.4.375

Grant, R. M. (1996b). Toward a knowledge-based theory of the firm. *Strategic Management Journal, 17*(S2), 109–122.

Grant, R. M. (2013). TOWARD A KNOWLEDGE-BASED THEORY OF THE FIRM, *17*, 109–122.

Grant, R. M., & Grant, R. M. (2014). Toward a Knowledge-Based Theory of the Firm. *Strategic Management Journal, 17*(S2), 109–122. https://doi.org/10.2307/2486994

Griffith, T. L., & Sawyer, J. E. (2010a). Multilevel knowledge and team performance, *1031*(May 2008), 1003–1031. https://doi.org/10.1002/job

Griffith, T. L., & Sawyer, J. E. (2010b). Multilevel knowledge and team performance. *Journal of Organizational Behavior, 31*(7), 1003–1031. https://doi.org/10.1002/job.660

Gubbins, C., Corrigan, S., Garavan, T. N., Connor, C. O., Leahy, D., Long, D., … Murphy, E. (2012). Evaluating a tacit knowledge sharing initiative: a case study. *European Journal of Training and Development, 36*(8), 827–847. https://doi.org/10.1108/03090591211263558

Gupta, A. K., & Govindarajan, V. (2000). Knowledge Flows Within Multinational

Corporations. *Strategic Management Journal Strat. Mgmt. J, 21*(August 1999), 473–496.

Haas, M. R., & Hansen, M. T. (2005). When using knowledge can hurt performance: The value of organizational capabilities in a management consulting company. *Strategic Management Journal, 26*(1), 1–24. https://doi.org/10.1002/smj.429

Hansen, M. (2002). Knowledge networks: Explaining effective knowledge sharing in multiunit companies. *Organization Science, 13*(3), 232–248. Retrieved from http://pubsonline.informs.org/doi/abs/10.1287/orsc.13.3.232.2771

Hansen, M. T. (2002). Knowledge Networks : Explaining Effective Knowledge Sharing in Multiunit Companies Knowledge Networks : Explaining Effective Knowledge Sharing in Multiunit Companies. *Organization Science, 13*(July 2014), 232–248. Retrieved from http://pubsonline.informs.org/doi/abs/10.1287/orsc.13.3.232.2771

Havlicek, L. L., & Peterson, N. L. (1976). Robustness of the Pearson Correlation against Violations of Assumptions. *Perceptual and Motor Skills, 43*(3_suppl), 1319–1334. https://doi.org/10.2466/pms.1976.43.3f.1319

Hosseini, S. M., & Akhavan, P. (2015). Determinants of Knowledge Sharing in Knowledge Networks : A Social Capital Perspective. *IUP Journal of Knowledge Management, 13*(1), 7–24. Retrieved from http://web.b.ebscohost.com/ehost/detail/detail?vid=7&sid=4816b46d-b2a0-400b-8366-586b80db9c13@sessionmgr110&hid=128&bdata=JnNpdGU9ZWhvc3QtbGl2ZQ==#AN=101858505&db=buh

Huang, M.-C., Chiu, Y.-P., & Lu, T.-C. (2013). Knowledge governance mechanisms and repatriate's knowledge sharing: the mediating roles of motivation and opportunity. *Journal of Knowledge Management, 17*(5), 677–694. https://doi.org/10.1108/JKM-01-2013-0048

Huggins, R., & Johnston, A. (2010). Knowledge flow and inter-firm networks: The influence of network resources, spatial proximity and firm size. *Entrepreneurship & Regional Development, 22*(5), 457–484. https://doi.org/10.1080/08985620903171350

Huggins, R., Johnston, A., & Stride, C. (2012). Knowledge networks and universities: locational and organisational aspects of knowledge transfer interactions. *Entrepreneurship and Regional Development, 24*(7–8), 475–502. https://doi.org/10.1080/08985626.2011.618192

Inkpen, A. C., & Dinur, A. (2013). Knowledge Management Processes Ventures and International Joint, *9*(4), 454–468.

Insch NancyDawley, David, G. S. M. (2008). Tacit Knowledge: A Refinement and Empirical Test of the Academic Tacit Knowledge Scale. *Journal of Psychology, 142*(6), 561–580. https://doi.org/10.3200/JRLP.142.6.561-580

Ipe, M. (2003). Knowledge Sharing in Organizations: A Conceptual Framework. *Human Resource Development Review, 2*(4), 337–359. https://doi.org/10.1177/1534484303257985

Jo, S. J., & Joo, B.-K. (Brian). (2011). Knowledge Sharing: The Influences of Learning Organization Culture, Organizational Commitment, and Organizational Citizenship Behaviors. *Journal of Leadership & Organizational Studies, 18*(3), 353–364. https://doi.org/10.1177/1548051811405208

Karkoulian, S., Messarra, L. C., & McCarthy, R. (2013). The intriguing art of knowledge management and its relation to learning organizations. *Journal of Knowledge Management, 17*(4), 511–526. https://doi.org/10.1108/JKM-03-2013-0102

Katsamakas, E. (2007). Knowledge processes and learning options in networks: Evidence from telecommunications. *Human Systems Management, 26*(3), 181–192. Retrieved from http://search.ebscohost.com/login.aspx?direct=true&db=buh&AN=26501607&site=ehost-live

Kendall, M., & Gibbons, J. D. R. (1990). Correlation methods. Oxford: Oxford University Press.

Kikoski, C. K., & Kikoski, J. F. (2004). *The inquiring organization: Tacit knowledge, conversation, and knowledge creation: Skills for 21st-century organizations.* Greenwood Publishing Group.

Kowalski, C. J. (1972). On the Effects of Non-Normality on the Distribution of the Sample Product-Moment Correlation Coefficient. *Journal of the Royal Statistical Society. Series C (Applied Statistics), 21*(1), 1–12. Retrieved from http://www.jstor.org/stable/2346598

Krogh, G. Von, Roos, J., & Slocum, K. (1994). An essay on corporate epistemology. *Strategic Management Journal, 15*(Summer), 53–71. https://doi.org/10.1002/smj.4250151005

Kumar, N. (2013). Managing reverse knowledge flow in multinational corporations. *Journal of Knowledge Management, 17*(5), 695–708. https://doi.org/10.1108/JKM-02-2013-0062

Lahtinen, J. (2013). Local social knowledge management: A case study of social learning and knowledge sharing across organizational boundaries. *Journal of Information Science, 39*(5), 661–675. https://doi.org/Doi 10.1177/0165551513481431

Lee, S., Suh, E., & Lee, M. (2014). Measuring the risk of knowledge drain in communities of practice. *Journal of Knowledge Management, 18*(2), 382–395. https://doi.org/10.1108/JKM-07-2013-0263

Lin, H.-F. (2007). Effects of extrinsic and intrinsic motivation on employee knowledge sharing intentions. *Journal of Information Science, 33*(2), 135–149. https://doi.org/10.1177/0165551506068174

Lorentzen, A. (2008). Knowledge networks in local and global space. *Entrepreneurship & Regional Development*, *20*(6), 533–545. https://doi.org/10.1080/08985620802462124

March, J. G. (1991). Exploration and Exploitation in Organizational Learning. *Organization Science*, *2*(1), 71–88. https://doi.org/10.1287/mnsc.43.7.934

Mariano, S., & Awazu, Y. (2016). Artifacts in knowledge management research: a systematic literature review and future research directions. *Journal of Knowledge Management*, *20*(6). https://doi.org/10.1108/JKM-05-2016-0199

Matusik, S. F., & Hill, C. W. L. (1998). The Utilization of Contingent Knowledge Creation , and Competitive Advantage. *Academy of Management Review*, *23*(4), 680–697. https://doi.org/10.5465/AMR.1998.1255633

McIver, D., Lengnick-Hall, C. A., Lengnick-Hall, M. L., & Ramachandran, I. (2013). Understanding work and knowledge management from a knowledgein-practice perspective. *Academy of Management Review*, *38*(4), 597–620. https://doi.org/10.5465/amr.2011.0266

Montague, W. P. (1925). *The ways of knowing or the methods of philosophy. The Muirhaed Library of Philosophy*. London: Compton Printing Works Ltd.

Mukaka, M. M. (2012). A guide to appropriate use of Correlation coefficient in medical research. *Malawi Medical Journal*, *24*(3), 69–71. https://doi.org/10.1016/j.cmpb.2016.01.020

Mura, M., Lettieri, E., Radaelli, G., & Spiller, N. (2013). Promoting professionals' innovative behaviour through knowledge sharing: the moderating role of social capital. *Journal of Knowledge Management*, *17*(4), 527–544. https://doi.org/10.1108/JKM-03-2013-0105

MySQL. (n.d.). Oracle Inc. Retrieved from http://www.mysql.com

MySQL, A. (2006). Mysql connector/j. Oracle Inc. Retrieved from http://scholar.google.com/scholar?hl=en&btnG=Search&q=intitle:MySQL+Connector+/ +J#0

Nakano, D., Muniz, J., & Batista, E. D. (2013). Engaging environments: tacit knowledge sharing on the shop floor. *Journal of Knowledge Management, 17*(2), 290–306. https://doi.org/10.1108/13673271311315222

Nasierowski, W., & Mikula, B. (2011). The Polish Culture-Social-Economic Features as a Basis to Create Knowledge-Based Organizations ' Culturs, *3*(1), 64–80.

Nerkar, A., & Paruchuri, S. (2005). Evolution of R&D capabilities: The role of knowledge networks within a firm. *Management Science, 51*(5), 771–785. https://doi.org/DOI 10.1287/mnsc.1040.0354

Neumann, E., & Prusak, L. (2007). Knowledge networks in the age of the Semantic Web. *Briefings in Bioinformatics, 8*(3), 141–149. https://doi.org/10.1093/bib/bbm013

Newell, A., & Simon, H. a. (1976). Computer science as empirical inquiry: symbols and search. *Communications of the ACM, 19*(3), 113–126. https://doi.org/10.1145/360018.360022

Newman, M. (2003). The structure and function of complex networks. *SIAM Review, 45*(2), 167–256. https://doi.org/10.1137/S003614450342480

Newman, M., Barabasi, A.-L., & Watts, D. J. (2006). *The structure and dynamics of networks.* Princeton University Press.

Nonaka, I. (1991). The knowledge-creating company. *Harvard Business Review*, (August). Retrieved from http://www3.uma.pt/filipejmsousa/ge/Nonaka, 1991.pdf

Nonaka, I. (1994). A dynamic theory of organizational knowledge creation. *Organization Science, 5*(1), 14–37. Retrieved from http://pubsonline.informs.org/doi/abs/10.1287/orsc.5.1.14

Nonaka, I. (1994). A Dynamic Theory of Organizational Knowledge Creation. *Organization Science*, *5*(1), 14–37. Retrieved from http://business.illinois.edu/josephm/BA504_Fall 2008/Uploaded in Nov 2007/Nonaka (1994).pdf

Nonaka, I. (2007). The Knowledge-Creating Company. *Harvard Business Review*, (August). https://doi.org/10.1016/S0969-4765(04)00066-9

Nonaka, I., & Takeuchi, H. (1995). *The Knowledge-Creating: How Japanese companies create the dynamics of innovation. Oxford University Press* (Vol. 3). New York: Oxford University Press. https://doi.org/10.1016/S0048-7333(97)80234-X

Panahi, S., Watson, J., & Partridge, H. (2013). Towards tacit knowledge sharing over social web tools. *Journal of Knowledge Management*, *17*(3), 379–397. https://doi.org/10.1108/JKM-11-2012-0364

Payne, J. W., Bettman, J. R., & Johnson, E. J. (1993). *The adaptive decision maker. The Adaptive Decision Maker* (Vol. 45). Cambridge University Press. https://doi.org/10.1057/jors.1994.133

Pemberton, J. D., & Stonehouse, G. H. (2000). Organisational learning and knowledge assets--an essential partnership. *Learning Organization, The*, *7*(4), 184–194. https://doi.org/10.1108/09696470010342351

Peng, H. (2013). Why and when do people hide knowledge? *Journal of Knowledge Management*, *17*(3), 398–415. https://doi.org/http://dx.doi.org/10.1108/JKM-12-2012-0380

Phelps, C., Heidl, R., & Wadhwa, A. (2012). *Knowledge, Networks, and Knowledge Networks: A Review and Research Agenda. Journal of Management* (Vol. 38).

https://doi.org/10.1177/0149206311432640

Plato, & Jowett, B. (2012). *Meno.*

Poh-Kam, W. (2000). Knowledge Creation Management: Issues and Challenges. *Asia Pacific Journal of Management, 17*(2), 193-200–200. https://doi.org/10.1023/A:1015845008209

Polanyi, M. (1966). *The Tacit Dimension.* University of Chicago Press.

Polanyi, M. (1967). *The tacit dimension.* The University of Chicago Press.

Prahalad, C. K., & Bettis, R. A. (1986). The dominant logic: A new linkage between diversity and performance. *Strategic Management Journal, 7*(6), 485–501. https://doi.org/10.1002/smj.4250070602

R Development Core Team. (2008). R: A Language and Environment for Statistical Computing. Vienna, Austria.

Ramírez, A. M., Morales, V. J. G., & Rojas, R. M. (2011). Knowledge Creation, Organizational Learning and Their Effects on Organizational Performance. *Žinių Kūrimas Ir Organizacijos Mokymasis Bei Jų Poveikis Organizacijos Veiklai., 22*(3), 309–318. https://doi.org/10.5755/j01.ee.22.3.521

Ribino, P., Augello, A., Lo Re, G., & Gaglio, S. (2011). A knowledge management and decision support model for enterprises. *Advances in Decision Sciences, 2011*, 1–16. https://doi.org/10.1155/2011/425820

Rodriguez, E., & Edwards, J. S. (2009). Applying knowledge management to enterprise risk management: Is there any value in using KM for ERM? *Journal of Risk Management in Financial Institutions, 2*(4), 427–437. Retrieved from http://search.ebscohost.com/login.aspx?direct=true&db=buh&AN=44369856&site=bsi-live

Rosvall, M., & Bergstrom, C. T. (2008). Maps of random walks on complex networks reveal community structure. *Proceedings of the National Academy of Sciences of the United States of America*, *105*(4), 1118–1123. https://doi.org/10.1073/pnas.0706851105

Rowley, J. (2007). The wisdom hierarchy: representations of the DIKW hierarchy. *Journal of Information Science*, *33*(2), 163–180. https://doi.org/10.1177/0165551506070706

Rünger, D. (2012). How sequence learning creates explicit knowledge: The role of response-stimulus interval. *Psychological Research*, *76*(5), 579–590. https://doi.org/10.1007/s00426-011-0367-y

Ryan, S., & O'Connor, R. V. (2013). Acquiring and Sharing tacit knowledge in software development teams: An empirical study. *Information and Software Technology*, *55*(9), 1614–1624. https://doi.org/10.1016/j.infsof.2013.02.013

Ryu, C., Kim, Y. J., Chaudhury, A., & Rao, H. R. (2005). Knowledge Acquisition via Three Learning Processes in Enterprise Information Portals: Learning-by-Investment, Learning-by-Doing, and Learning-from-Others. *MIS Quarterly*, *29*(2), 245–278. Retrieved from http://misnt.indstate.edu/harper/s105/cb/sohmarcusgoh.pdf

Seidler-de Alwis, R., & Hartmann, E. (2008). The use of tacit knowledge within innovative companies: knowledge management in innovative enterprises. *Journal of Knowledge Management*, *12*(1), 133–147. https://doi.org/10.1108/13673270810852449

Shank, R. C., & Abelson, R. P. (1977). *Scripts, Plans, Goals, and Understanding: An Inquiry Into Human Knowledge Structures*. New Jersey: Lawrence Erlbaum Associates, Publishers.

Sieloff, C. G. (1999). "If only HP knew what HP knows": the roots of knowledge management at Hewlett-Packard. *Journal of Knowledge Management*, *3*, 47–53. https://doi.org/10.1108/13673279910259385

Simon, H. A. (1950). *Administrative behaviour. Australian Journal of Public Administration* (Vol. 9). Wiley Online Library. https://doi.org/10.1111/j.1467-8500.1950.tb01679.x

Simon, H. A. (1999). Bounded Rationality and Organizational Learning. *Reflections, 1*(2), 17/27. https://doi.org/10.1287/orsc.2.1.125

Simonin, B. L. (1999). *Ambiguity and the Process of Knowledge Transfer in Strategic Alliances. Igarss 2014* (Vol. 20). Wiley Online Library. https://doi.org/10.1007/s13398-014-0173-7.2

Small, C. T., & Sage, A. P. (2006). Knowledge management and knowledge sharing : A review. *Information Knowledge Systems Management, 5*(3), 153–169. https://doi.org/10.1017/S0954579409000157

Smith, H. A., & McKeen, J. D. (2004). Creating and facilitating communities of practice. In *Handbook on Knowledge Management 1* (pp. 393–407). Springer.

Song, J. H., & Chermack, T. J. (2008). A theoretical approach to the organizational knowledge formation process: Integrating the concepts of individual learning and learning organization culture. *Human Resource Development Review, 7*(4), 424–442. https://doi.org/10.1177/1534484308324983

Spencer, H., & Wiseman, D. (n.d.). University of Toronto, Department of Zoology (UTZOO) Usenet Archive. Retrieved from https://archive.org/details/utzoo-wiseman-usenet-archive

Spender, J.-C. (2008). Organizational Learning and Knowledge Management: Whence and Whither? *Management Learning, 39*(2), 159–176. https://doi.org/10.1177/1350507607087582

Staples, D., Greenaway, K., & McKeen, J. (2001). Opportunities for research about managing the knowledge-based enterprise. *International Journal of Management Reviews, 3*(1), 1–

20. https://doi.org/10.1111/1468-2370.00051

Štorga, M., Mostashari, A., & Stankovic, T. (2013). Visualisation of the organisation knowledge structure evolution. *Journal of Knowledge Management, 17*(5), 724–740. https://doi.org/10.1108/JKM-02-2013-0058

Stough, S., Eom, S., & Buckenmyer, J. (2000). Virtual teaming: a strategy for moving your organization into the new millennium. *Industrial Management & Data Systems, 100*(8), 370–378. https://doi.org/10.1108/02635570010353857

Tan, B. C. Y. (2005). v Hid ,! CwIlV secalssuE Knowledge Contributing Repositories : Knowledge to Electronic An Empirical, *29*(1), 113–143.

Tanhua-Piiroinen, E., & Sommers-Piiroinen, J. (2013). Knowledge Sharing Cultures in Finance and Insurance Companies - Needs for Improving Informal Collaborative e-Learning. *International Journal of Advanced Corporate Learning, 6*(2), 36–39. https://doi.org/10.3991/ijac.v6i2.2983

Tao, X., Li, Y., Zhong, N., & Nayak, R. (2008). An ontology-based framework for knowledge retrieval. *Proceedings - 2008 IEEE/WIC/ACM International Conference on Web Intelligence, WI 2008*, 510–517. https://doi.org/10.1109/WIIAT.2008.226

Thorburn, L. (2000). Knowledge management , research spinoffs commercialization of R & D in Australia. *Asia Pacific Journal of Management, 17*, 257–275. https://doi.org/10.1023/A:1015861625956

Tuomi, I. (1999). Data is more than knowledge: implications of the reversed knowledge hierarchy for knowledge management and organizational memory. *J. Manage. Inf. Syst., 16*(3), 103–117.

Ubuntu. (n.d.). Canonical. Retrieved from http://www.ubuntu.com

van den Berg, H. a. (2013). Three shapes of organisational knowledge. *Journal of Knowledge Management*, *17*(2), 159–174. https://doi.org/10.1108/13673271311315141

Van Rijsbergen, C. J., Robertson, S. E., & Porter, M. F. (1980). New models in probabilistic information retrieval. *British Library Research and Development Report*, *5587*(5587), 123. Retrieved from http://tartarus.org/~martin/PorterStemmer

Varela, F., Thompson, E., & Rosch, E. (1992). *The embodied mind*. Cambridge, MA: The MIT Press.

Walsh, J. P., & Ungson, G. R. (1991). Organizational memory. *Academy of Management Review*, *16*(1), 57–91.

Walsh, J., & Ungson, G. (1991). Organizational memory. *Academy of Management Review*, *16*(I). Retrieved from http://amr.aom.org/content/16/1/57.short

Wang, C., Rodan, S., Fruin, M., & Xu, X. (2014). Knowledge networks, collaboration networks, and exploratory innovation. In *Academy of Management Journal* (Vol. 57, pp. 484–514). https://doi.org/10.5465/amj.2011.0917

Wegner, D. M., Erber, R., & Raymond, P. (1991). Transactive memory in close relationships. *Journal of Personality and Social Psychology*, *61*(6), 923–929. https://doi.org/10.1037/0022-3514.61.6.923

Wenger, E. (1998). *Communities of practice: Learning, meaning, and identity*. Cambridge university press.

Wenger, E. (1999). *Communities of practice: Learning, meaning, and identity*. Cambridge university press.

Wenger, E. C., & Snyder, W. M. (2000). Communities of practice: The organizational frontier. *Harvard Business Review*, *78*(1), 139–146.

Whyte, G., Classen, S., Authors, F., Whyte, G., & Classen, S. (2012). Using storytelling to elicit tacit knowledge from SMEs. *Journal of Knowledge Management, 16*(6), 950–962. https://doi.org/10.1108/13673271211276218

WIL. (16AD). WildFly Application Server. Redhat Inc. Retrieved from https://docs.jboss.org/author/display/WFLY10/Documentation

Yang, J. T. (2009). Individual attitudes to learning and sharing individual and organisational knowledge in the hospitality industry. *Service Industries Journal, 29*(12), 1723–1743. https://doi.org/10.1080/02642060902793490

Zelleny, M. (2002). Knowledge of enterprise: knowledge management or knowledge technology? *International Journal of Information Technology & Decision Making, 1*(2), 181–207.

Zeller, R. A., & Levine, Z. H. (1974). The Effects of Violating the Normality Assumption Underlying r. *Sociological Methods & Research, 2*(4), 511–519. https://doi.org/10.1177/004912417400200406

Appendix A   System Architecture

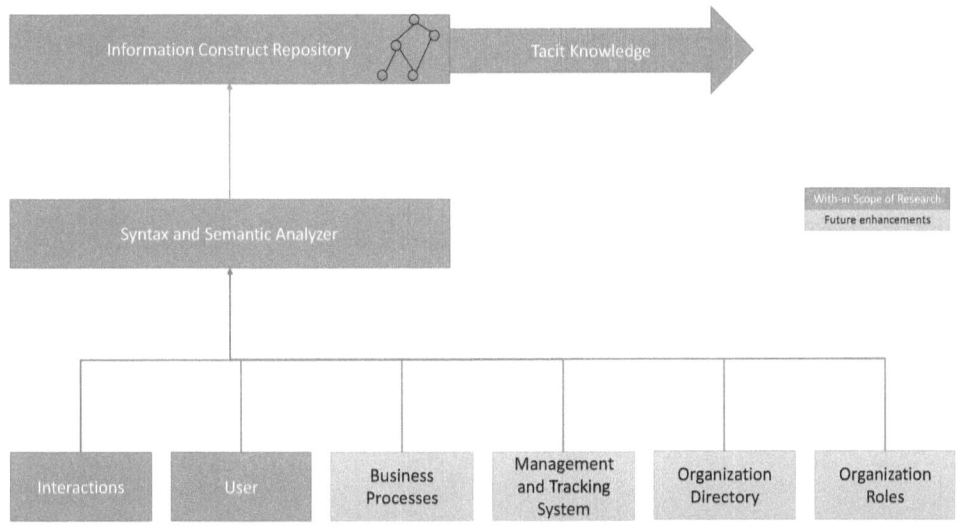

Figure 48 Context of the system; functional blocks that were built or built in future

In the context of an enterprise, multiple sources need to be consumed to capture the tacit knowledge. Figure 48 defines the context in which tacit knowledge can be managed within an organization or group. Any system that is built to track and manage tacit knowledge takes its inputs from multiple sources. These sources include organization roles, organization social graph, interactions among individuals within the organization and organization processes. These sources are analyzed semantically and are used to build maps related to knowledge, skills, interests, etc.

Figure 49 depicts the architecture of the system that we build to process and communication repository of the communities.

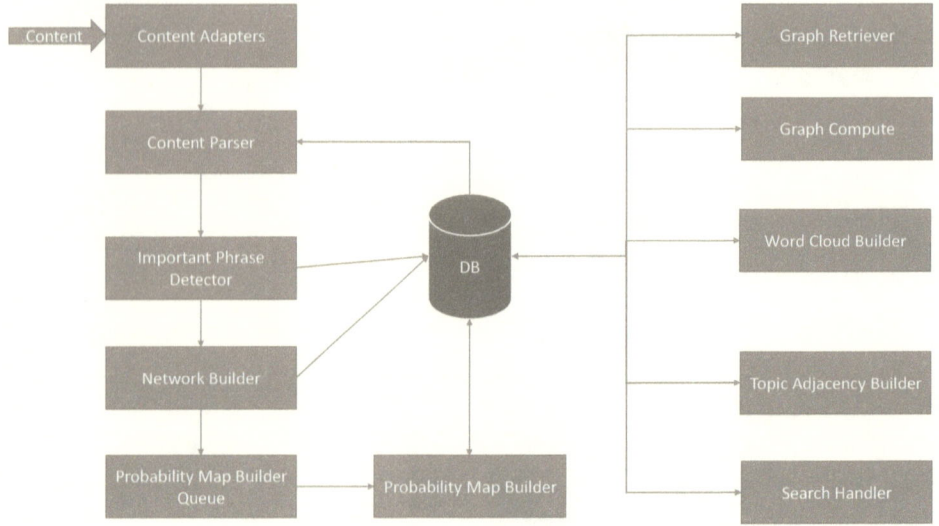

Figure 49 System Block Diagram; system modules that were implements as part of research

Following are the major modules of the system.

1. **Content Adapters** – Content adapter modules provide a mechanism to handle different types of content received by the system. For example, there is a minor difference in the way email messages are handled versus the Usenet messages. We need to create a content adapter for each type of content, and that adapter handles nuances of each type of content.

2. **Content Parsers** – Once the adapter has converted the content to a format that is common, the parser kicks in. The primary job of the parser is to handle textual parts of the content.

3. **Important Phrase Detector** – Once the contents are harmonized, the important phrase detector goes through the textual part of the content and generates the list of important phrases.

4. **Network Builder** – The next step in the process is to generate the data that contributes to the building of network. For each message received, this module generates nodes and edges relevant to that messages.

5. **Probability Map Queue** – Since probability map cannot be built with a single message, the messages are inserted into a queue from where they are periodically processed.

6. **Probability Map Builder** – This module generates the probability map of all the important phrases detected. This information is used later for adjacency calculation.

7. **Graph Retriever** – This module generates a graphml file from the computed data that is stored in the database. These graphml files are used to generate metrics related to the graph. We could generate graphs on per topic or for a period.

8. **Graph Compute** – This module generates all the metrics like betweenness, indegree, outdegree, community detection.

9. **Word Cloud Builder** – This module uses the probability map to generate the word cloud. It can generate word cloud on a per topic basis, for a period, for an individual.

10. **Topic Adjacency Builder** – This module generates topic adjacency for each phrase by using earlier generated probability map for further calculation.

11. **Search Handler** – This module handles all the search requests.

Formal System Architecture

In the formal architecture, we have merged blocks that were performing homogeneous functionality. Let's break down these blocks one by one.

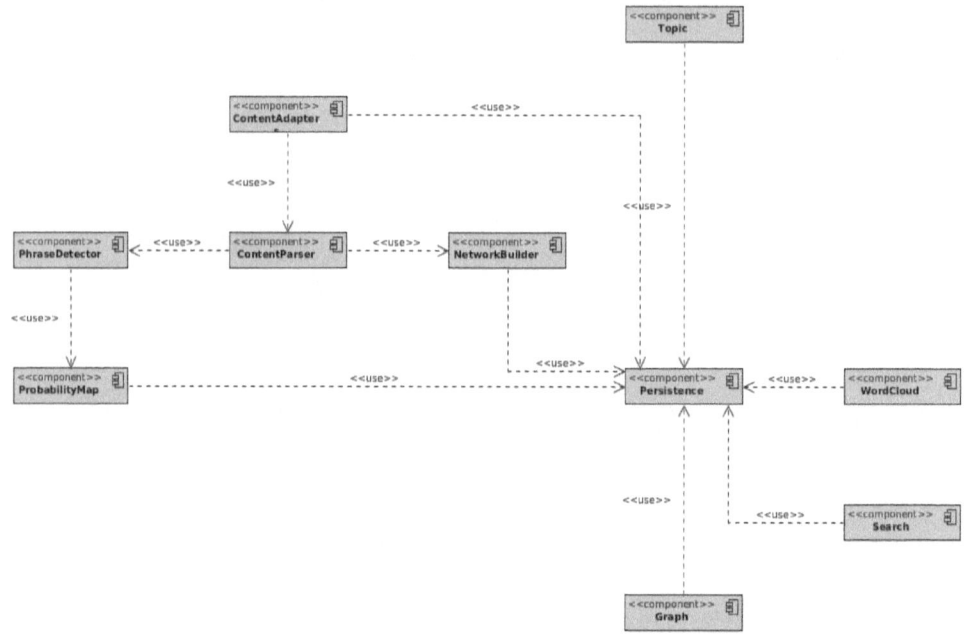

Figure 50 Top Level Architecture; Implementation components

Figure 50 defines the top level formal architecture for the system. The formal architecture is very like block diagram in Figure 49.

Dynamic Model

New Message Received

The system receives a new message; the message could be any of supported content types. It could be a simple text-based message or multimedia message. Figure 51 describes the steps the message needs to go through for complete processing. We use a standard factory pattern for having an extensible design where we can add new types without much problem(Ellis, Stylos, & Myers, 2007).

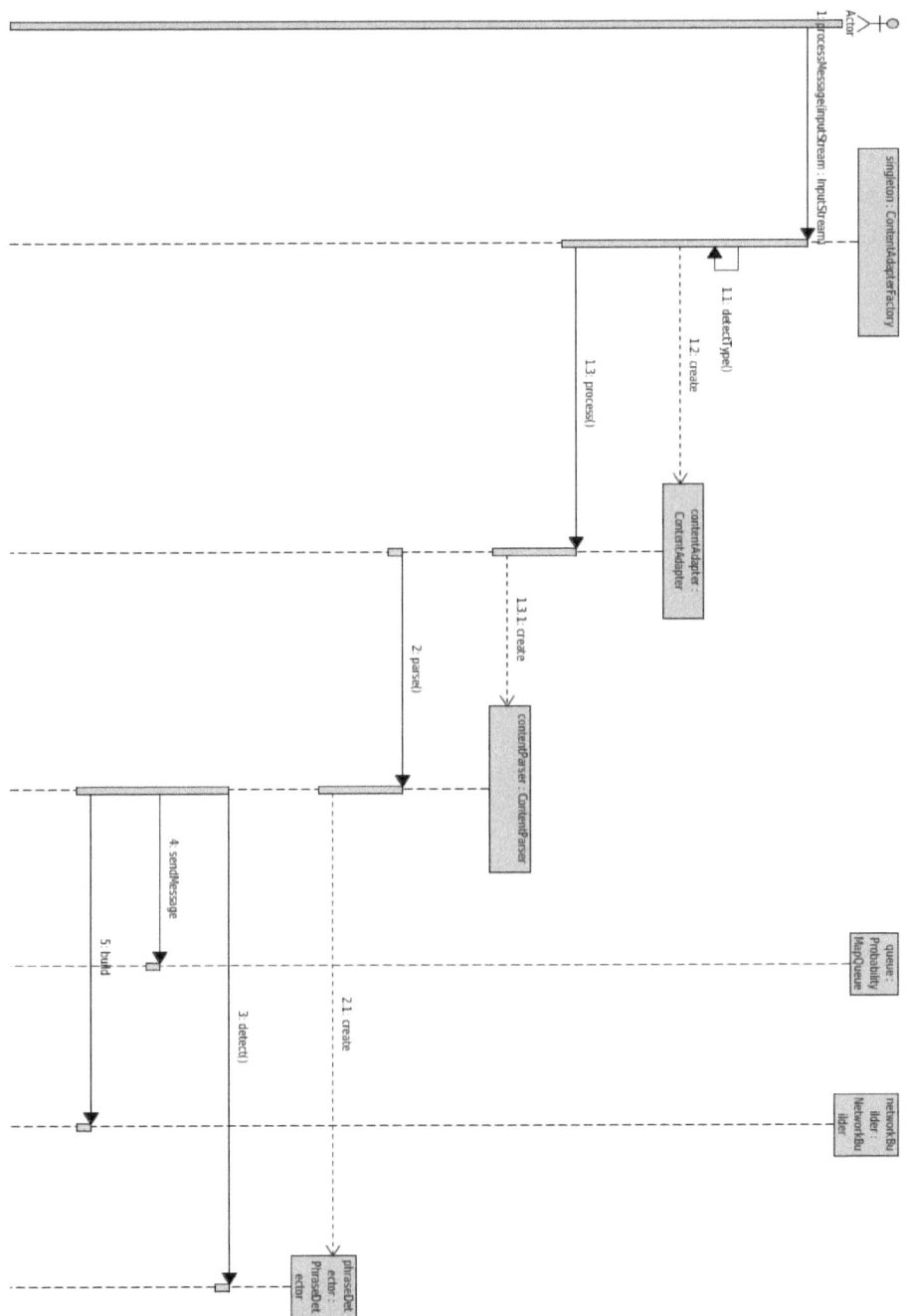

Figure 51 A new message is received; how a new message traverses through the system

147

Build a single edge

When the system parses a single message, it results in the creation of one or many edges in the network. Figure 52 describes the flow that is followed to create a single edge from a set of the sender, receiver, topic, and weight.

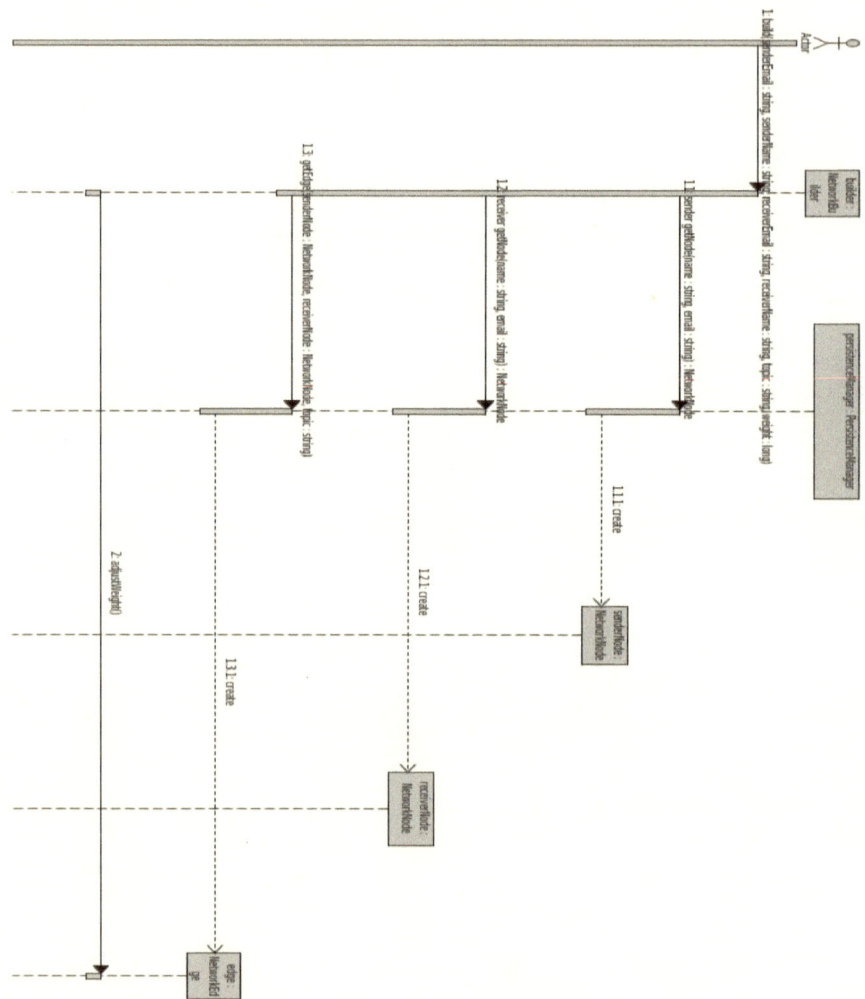

Figure 52 Build a single network edge

Create or retrieve a single node

Figure 52 references getNode which is detailed in Figure 53. The system first looks for an existing node in the repository with matching email address. If the node does not exist, then a new node is created and returned. If the node already exists, then it is retrieved and returned.

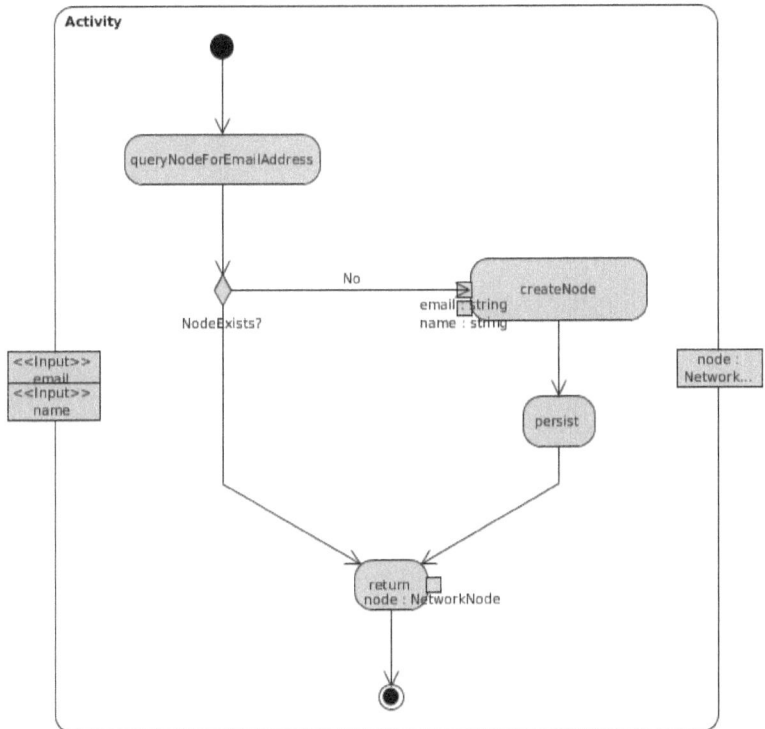

Figure 53 getNode Activity Diagram

Parsing the text

Figure 54 describes the steps that are followed for parsing the textual content. The processing of textual content follows through the following process.

- The text is tokenized using an Indo-European Tokenizer(Alias-i, 2008).

- Next step eliminates all non-alphanumeric phrases

- We convert the text to lowercase

- We run it through porter stemmer.

149

- We generate NGRAMs from the body of text with length = 2

These NGRAMs are basic building blocks for probability map calculation and other computation related to adjacency calculation.

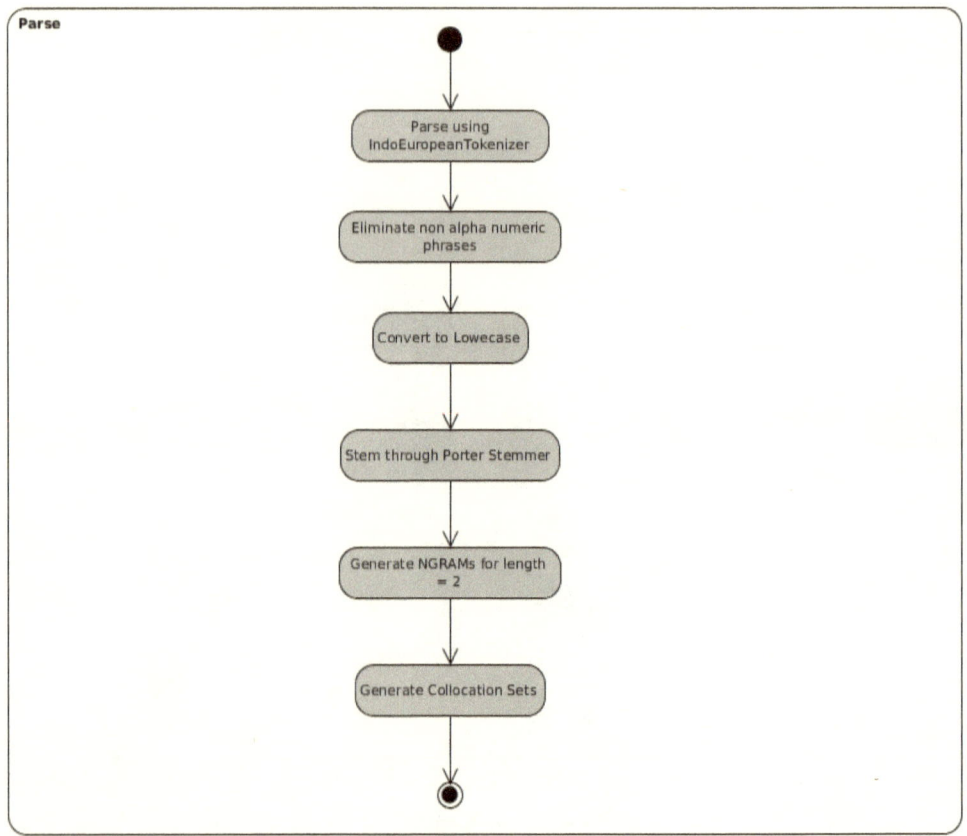

Figure 54 Parsing the text

Processing Message from Queue

Since the generation of adjacency data and graphs is a time-consuming process, most of this processing is done on an offline basis. The initial request inserts and message into the queue and the queue messages are processed on a periodic basis helping us in running these at times when the system is not overloaded.

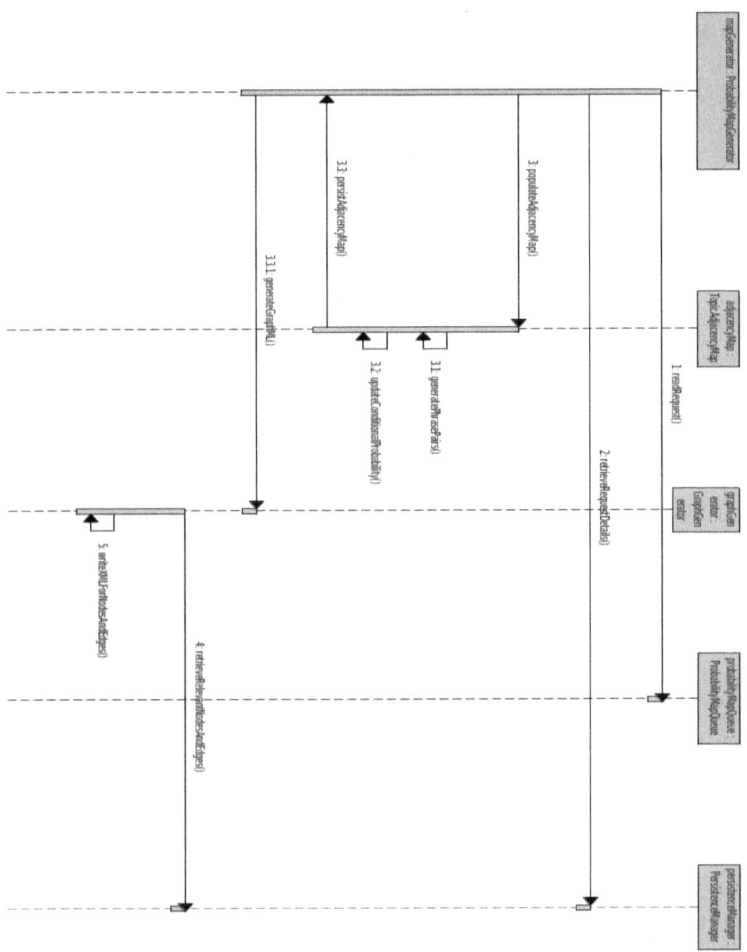

Figure 55 Offline Request Processing

Appendix B   Survey Questionnaire

In this section, we present an exact reproduction of Google Form Survey that was administered as part of the research. It was disseminated through email groups, LinkedIn groups, and other social media. The questionnaire was sent to 50 email addresses of colleagues and friends in the industry, an alumni group for Post Graduate Program in Software Enterprise Management consisting of 96 members, alumni group of Government Engineering College Bhopal consisting of 32 members, and employees of Hyderabad office of Tech Mahindra. Most of the audience has been employed in the knowledge economy. Of the total 146 responses received, 86 were from MNCs, 47 were from Private Sector, 13 were from Others. There were no responses received from PSU and Government organizations. Regarding roles, 45 responses were received from managers, ten responses were received from managers of managers, and 91 responses were received from individual contributors.

How well do we manage knowledge in organizations?

* Required

1. **What is your employer?** * Please categorize your employer. *Mark only one oval.*

   ◯ MNC
   ◯
   ◯ Private Sector
   ◯
   ◯ PSU

Government Others

2. Please categorize your role * *Mark only one oval.*

   ◯   I am an individual contributor   *Skip to question 19.*
   ◯
   ◯   I am a manager   Skip to question 11.

   I manage many managers   *Skip to question 3.*

Stop filling out this form.

Senior Managers

3. Are you considered a system and process expert in your organization? * *Mark only one oval.*

○   Yes   Skip to question 6.

○

No   Skip to question 8.

Senior Manager General

4. How often did you feel the need to contact a subject matter expert who is not part of your team/group? *

Do you face situations where getting in touch with an individual in some other team/group would be the best way to solve the problem/situation that you are facing?.. *Mark only one oval.*

	1	2	3	4	5	6	7	
Never	○	○	○	○	○	○	○	Very Often

5 How often did you feel the need to contact a subject matter expert who has left your team/group/company? *

Do you face situations when you feel the need to contact an ex-team member because he is the best and fastest way to get the information that you need to solve the problem/situation at hand?.

Mark only one oval.

	1	2	3	4	5	6	7	
Never	○	○	○	○	○	○	○	Very Often

Skip to question 23.

Senior Manager System and Process Expertise

6. How often are you requested by your peers regarding some help with organization processes and systems? *

Do you have situations when you peers may ask you help with wading through the organization bureaucracy? This is generally about the help that can't be found in organization documentation. *Mark only one oval.*

	1	2	3	4	5	6	7	
Never	○	○	○	○	○	○	○	Often

7. How often are members in your staff able to help others when your help is sought with organization processes and systems? *

When your help is sought, are there scenarios when you can ask somebody from your staff to help them?. *Mark only one oval.*

	1	2	3	4	5	6	7	
Never	○	○	○	○	○	○	○	Often

Skip to question 8.

Senior Manager choice of staff

8. **Do you have people in your staff who are considered organization process and system experts?** * Are there individuals in you staff whose help is requested in trying to understand organization systems and processes?

Mark only one oval.

○ Yes Skip to question 9.

○

No Skip to question 4.

Senior Managers with organization and system expert staff

Never	○	○	○	○	○	○	○	Often

9. How often are you contacted by other groups for help in organization systems and process from one of your staff members? *Mark only one oval.*

      1   2   3   4   5   6   7

10 (Optional)When any help is requested, how often do you think that you have multiple team members in your staff who can help with that problem/situation?

If the person in your staff whose help is requested can not help for some reason, do you have alternative individuals? for example individual who work in his team, who can help out in the given situation. *Mark only one oval.*

      1   2   3   4   5   6   7

Never ◯ ◯ ◯ ◯ ◯ ◯ ◯ Often

Skip to question 4.

People Manager

11. Are you considered a system and process expert in your organization? * *Mark only one oval.*

◯   Yes   Skip to question 17.

◯

    No   Skip to question 12.

People Manager General

12. How often did you feel the need to contact a subject matter expert who is not part of your team/group? *

Do you face situations where getting in touch with an individual in some other team/group would be the best way to solve the problem/situation that you are facing?.. *Mark only one oval.*

      1   2   3   4   5   6   7

Never ◯ ◯ ◯ ◯ ◯ ◯ ◯ Very Often

13. How often did you feel the need to contact a subject matter expert who has left your team/group/company? *

Do you face situations when you feel the need to contact an ex-team member because he is the best and fastest way to get the information that you need to solve the problem/situation at hand.

Mark only one oval.

      1   2   3   4   5   6   7

Never ◯ ◯ ◯ ◯ ◯ ◯ ◯ Very Often

14. Do you have individuals in your team who are considered subject matter experts across the organization?

Mark only one oval.

◯   Yes   Skip to question 15.

◯   No   Skip to question 23.

People Manager with Subject Matter Expert team members
15 How often are you requested by other teams/groups for help from a team member who is a subject matter expert? *

Do you have an individual in your team who is considered a subject matter expert and people from other groups/teams ask for help from him? *Mark only one oval.*

      1   2   3   4   5   6   7

Never ◯ ◯ ◯ ◯ ◯ ◯ ◯ Often

16. **How often the help of a subject matter can be replaced by appropriate documents and videos?** * Do you think that when other groups ask for the help of subject matter expert in your team, many times appropriate documents and videos could be good substitute? *Mark only one oval.*

	1	2	3	4	5	6	7	
Never	◯	◯	◯	◯	◯	◯	◯	Often

Skip to question 23.

People Manager Process and Systems Expertise

17. How often are you contacted by other people for help related to organization processes and systems? *

Organization processes and systems can sometime be tricky and people go to other individuals for help in using these processes and systems? *Mark only one oval.*

	1	2	3	4	5	6	7	
Never	◯	◯	◯	◯	◯	◯	◯	Very Often

18. How often are members in your team able to help others when your help is sought with organization processes and systems? *

When your help is sought, are there scenarios when you can ask somebody from your team to help them? *Mark only one oval.*

	1	2	3	4	5	6	7	
Never	◯	◯	◯	◯	◯	◯	◯	Very Often

Skip to question 12.

Individual Contributor

19. Are you considered a subject matter expert in the organization?

Are you considered an expert at something for which people seek your advice and help even if they are not your team members and do not work with you on a day to day basis? *Mark only one oval.*

○   Yes   Skip to question 21.

○

   No   Skip to question 20.

Individual Contributor General

20. How often do you feel the need of contacting a subject matter expert for the problem that you are/were working on? *

Do you ever feel that if you knew somebody with the knowledge, your time and effort could be saved and you can accomplish your tasks easily? The knowledge that you are looking for is very hard to find in the books or web. *Mark only one oval.*

	1	2	3	4	5	6	7	
Never	○	○	○	○	○	○	○	Very Often

Skip to question 23.

Individual Contributor subject matter expertise

21. How often are you contacted by other people/group regarding the subject matter that you are an expert in? *

Do other groups contact you about problems in your area of expertise? *Mark only one oval.*

	1	2	3	4	5	6	7	
Never	○	○	○	○	○	○	○	Very Often

22. How often have you been contacted by somebody in your previous group/employer?

* Do the groups where you previously worked every felt the need to consult you? *Mark only one oval.*

	1	2	3	4	5	6	7	
Never	○	○	○	○	○	○	○	Very Often

Skip to question 20.

Inaccurate	○	○	○	○	○	○	○	Accurate

Knowledge Management Systems

23. How useful do you find your organization's knowledge management system? * *Mark only one oval.*

	1	2	3	4	5	6	7	
Useless	○	○	○	○	○	○	○	Very useful

24.

How accurate is the information contained within the knowledge management system? * *Mark only one oval.*

1	2	3	4	5	6	7

25. How easy is the KM system to use? * *Mark only one oval.*

	1	2	3	4	5	6	7	
Difficult	○	○	○	○	○	○	○	Very easy

26. How well integrate is the KM system with rest of the tools that you use? * *Mark only one oval.*

1	2	3	4	5	6	7

Not integrated at all ⬭ ⬭ ⬭ ⬭ ⬭ ⬭ ⬭ Well integrated

27. What are the main impediments in your view towards the effective usage of KM system?

_____
_____
_____
_____
_____

28. What are the best things of the KM system?

_____
_____
_____
_____
_____

Skip to question 29.

Confirmation Page

29. Please leave your email address if you want to know the result of this survey.

_____

Appendix C   Literature Survey Details

We perform a literature review guided by following questions.

1. How do communities of practice evolve?

2. How does knowledge evolve in communities and individuals?

3. Why knowledge and not information or data?

4. How to get to tacit knowledge?

5. How to identify and differentiate communities and groups?

Based on these questions, we identify a set of keywords and search across EBSCO, Google Scholar, and Emerald. We eliminate obvious literature which doesn't fit our context by reading the title and abstract. After selecting the keywords and searching through the databases, we are left with the following number of documents. Table 14 contains the count of documents after elimination process that are considered for further analysis.

Table 14 Count of relevant documents

Keyword	Relevant document count
communities of practice	84
tacit knowledge	176
explicit knowledge	123
knowledge transfer	136
knowledge sharing	168
codified	99

Table 15 lists the frequency of chosen keywords in each of the papers selected for the literature review. We first go through the papers with higher keyword frequency and the follow up on relevant references in those papers.

Table 15 Keyword frequency in documents

Paper	Codification	Communities of Practice	Explicit Knowledge	Network Knowledge	Knowledge Sharing	Knowledge Transfer	Tacit Knowledge
(Haas & Hansen, 2005)	69	0	0	0	0	0	0
(van den Berg, 2013)	64	0	0	0	0	0	32
(M. T. Hansen, 2002)	24	0	0	0	0	0	0
(Bosua, 2013)	15	0	0	0	0	0	0
(Coff et al., 2006)	14	0	0	0	0	11	57
(Zelleny, 2002)	11	0	0	0	0	0	10
(Ardichvili et al., 2003)	0	64	0	0	79	0	0
(Etienne Wenger, 1999)	0	47	0	0	0	0	0
(Smith & McKeen, 2004)	0	22	0	0	0	0	0
(Lee et al., 2014)	0	18	0	12	18	0	13
(Neumann & Prusak, 2007)	0	17	0	15	0	0	0

Paper	Practice Codification	Communities of Practice	Explicit Knowledge	Network Knowledge	Knowledge Sharing	Knowledge Transfer	Tacit Knowledge
(Griffith & Sawyer, 2010b)	0	0	92	0	41	0	97
(Becerra et al., 2008)	0	0	88	0	0	54	71
(S. Cook & Brown, 1999)	0	0	32	0	0	0	61
(Ramírez et al., 2011)	0	0	21	0	0	0	11
(Rünger, 2012)	0	0	21	0	0	0	0
(Alavi & Leidner, 2001)	0	0	19	0	0	23	21
(Seidler-de Alwis & Hartmann, 2008)	0	0	17	0	11	16	137
(Alipour, Idris, & Karimi, 2011)	0	0	17	0	0	0	13
(Inkpen & Dinur, 2013)	0	0	15	0	0	18	14
(Connell, Klein, & Powell, 2003)	0	0	12	0	0	18	29
(Ikujiro Nonaka, 2007)	0	0	11	0	0	0	19
(Phelps et al., 2012)	0	0	0	136	19	47	14
(Wang, Rodan, Fruin, & Xu, 2014)	0	0	0	89	0	0	0

Paper	Practice Codification	Communities of Knowledge	Explicit Knowledge	Network Knowledge	Knowledge Sharing	Knowledge Transfer	Tacit Knowledge
(Bush, 2005)	0	0	0	46	0	0	0
(Huggins & Johnston, 2010)	0	0	0	43	0	0	0
(Hosseini & Akhavan, 2015)	0	0	0	29	68	0	0
(Nerkar & Paruchuri, 2005)	0	0	0	29	0	0	0
(Huggins, Johnston, & Stride, 2012)	0	0	0	27	0	34	0
(Lorentzen, 2008)	0	0	0	16	0	0	13
(Štorga, Mostashari, & Stankovic, 2013)	0	0	0	15	0	0	0
(Huang, Chiu, & Lu, 2013)	0	0	0	0	205	33	0
(Lin, 2007)	0	0	0	0	133	0	0
(Ipe, 2003)	0	0	0	0	104	0	0
(Panahi, Watson, & Partridge, 2013)	0	0	0	0	102	0	204
(Bakker et al., 2006)	0	0	0	0	74	0	0

Paper	Practice Codification	Communities of	Knowledge Explicit	Network Knowledge	Sharing Knowledge	Transfer Knowledge	Knowledge	Tacit Knowledge
(Small & Sage, 2006)	0	0	0	0	66	0	16	
(Yang, 2009)	0	0	0	0	60	0	0	
(Mura et al., 2013)	0	0	0	0	53	0	0	
(Nakano et al., 2013)	0	0	0	0	43	0	37	
(de Vries et al., 2006)	0	0	0	0	40	0	0	
(Lahtinen, 2013)	0	0	0	0	39	0	0	
(Jo & Joo, 2011)	0	0	0	0	33	0	0	
(Rodriguez & Edwards, 2009)	0	0	0	0	23	0	0	
(Tanhua-Piiroinen & Sommers-Piiroinen, 2013)	0	0	0	0	22	0	0	
(Peng, 2013)	0	0	0	0	21	0	0	
(Tan, 2005)	0	0	0	0	19	0	0	
(Karkoulian, Messarra, & McCarthy, 2013)	0	0	0	0	17	0	0	
(Ryan & O'Connor, 2013)	0	0	0	0	14	0	132	

Paper	Practice Codification	Communities of Knowledge	Explicit Knowledge	Network Knowledge	Sharing Knowledge	Transfer Knowledge	Knowledge	Tacit Knowledge
(Chau, Maurer, & Melnik, 2003)	0	0	0	0	12	0	0	
(Mariano & Awazu, 2016)	0	0	0	0	11	0	0	
(Chatzkel, 2004)	0	0	0	0	10	0	0	
(Simonin, 1999)	0	0	0	0	0	58	0	
(Fang, Jiang, Makino, & Beamish, 2010)	0	0	0	0	0	54	0	
(Kumar, 2013)	0	0	0	0	0	53	0	
(Gupta & Govindarajan, 2000)	0	0	0	0	0	23	0	
(Staples, Greenaway, & McKeen, 2001)	0	0	0	0	0	14	0	
(Burkhard, Hill, & Venkatsubramanyan, 2011)	0	0	0	0	0	12	0	
(Argote, 2005)	0	0	0	0	0	11	0	
(Gable, 2005)	0	0	0	0	0	11	0	
(Ryu, Kim, Chaudhury, & Rao, 2005)	0	0	0	0	0	10	0	

Paper	Codification	Practice	Communities of	Knowledge	Explicit	Network	Knowledge	Sharing	Knowledge	Transfer	Knowledge	Tacit Knowledge
(Insch NancyDawley, David, 2008)	0	0	0	0	0	0	99
(Gubbins et al., 2012)	0	0	0	0	0	0	91
(Avasthi, Vinay & Dey, 2015)	0	0	0	0	0	0	53
(Whyte, Classen, Authors, Whyte, & Classen, 2012)	0	0	0	0	0	0	32
(Thorburn, 2000)	0	0	0	0	0	0	25
(Akbar, 2003)	0	0	0	0	0	0	21
(McIver, Lengnick-Hall, Lengnick-Hall, & Ramachandran, 2013)	0	0	0	0	0	0	18
(Song & Chermack, 2008)	0	0	0	0	0	0	14
(Grant, 1996a)	0	0	0	0	0	0	14
(Poh-Kam, 2000)	0	0	0	0	0	0	13
(Grant & Grant, 2014)	0	0	0	0	0	0	12
(Katsamakas, 2007)	0	0	0	0	0	0	10
(Spender, 2008)	0	0	0	0	0	0	10

Appendix D   Phrase Representation Results

Table 16 contains the contribution of each of the phrases to the cliques that are formed. We can observe that most phrases belong to a single group.

Table 16 Detailed Phrase representation results

	Group 1	Group 2	Group 3	Group 4	Group 5	Group 6	Group 7	Group Represented
acceler		5.14%					1.17%	2
acceller		0.24%						1
ago		0.96%						1
akaa			0.62%					1
along		0.96%						1
alpha			0.47%					1
alphag			0.31%					1
am			0.70%					1
analog			1.63%					1
anisotropi							1.10%	1
appeal		0.96%						1
appl			0.62%					1
approch		0.72%					0.21%	2
arpa		0.48%	0.23%		4.62%		0.21%	4
articl			1.40%				14.44%	2
aspers			0.70%					1
author		0.96%						1
back								0
background							4.75%	1
ball		0.48%						1
bandwidth			0.47%					1
base			0.23%					1
behind		0.72%						1
believ			0.93%					1
beta			0.54%				2.48%	2

word							count
bill			0.47%				1
blackbodi						0.83%	1
both			0.70%				1
brain			0.31%				1
breath							0
bsw						0.21%	1
candi							0
candl					4.62%		1
case			0.70%				1
cast			0.70%				1
causal		0.24%	0.23%				2
cbmvax			0.47%				1
cbosgd						0.21%	1
cerenkov			0.47%				1
chang		0.96%	0.62%				2
channel		0.32%					1
claim		0.48%					1
clock		0.96%	0.47%				2
columbia			1.86%				1
comment						0.83%	1
comput			1.86%				1
consid		3.85%					1
constant		1.20%	0.93%				2
core	100.00%						1
csnet			0.47%				1
curv						0.83%	1
dad							0
davi					9.23%		1
defeat						14.44%	1
definit			0.62%				1
design			0.47%				1
determin			0.23%				1
determinist			0.47%				1
diagram		2.17%					1

dice		0.23%					1
did	0.32%						1
differ	0.32%	0.23%					2
digit		0.78%					1
dilat	0.72%						1
dimensionless		0.39%					1
discuss		1.16%					1
distribut		0.23%					1
do	0.40%	1.78%					2
doe	1.28%						1
doesn	0.40%						1
don		0.93%					1
dream							0
dt						1.51%	1
dust						1.10%	1
earth		0.23%					1
edg						0.21%	1
effect	0.96%						1
either	0.72%						1
element						0.83%	1
elev	1.52%						1
ethan						2.06%	1
even		0.93%					1
event	3.45%	0.23%					2
exampl		0.23%					1
experi	0.80%	0.78%					2
explan	0.96%						1
exponenti		0.23%					1
ey		1.16%					1
face						0.41%	1
falsifi		0.47%					1
feel							0
few						0.83%	1
field	0.24%						1

filler						14.44%	1
filter		0.47%					1
fine	0.24%						1
first	0.24%	0.23%					2
flame					6.15%		1
flash		2.87%					1
forc	0.24%					0.21%	2
frame						0.55%	1
galaxi						1.10%	1
gathman	0.24%						1
gener		0.78%					1
get							0
give		0.39%					1
go		0.23%					1
got	0.24%						1
graviti	0.48%					0.21%	2
greg							0
gwyn						1.03%	1
hand				5.56%			1
happen							0
harnad		0.70%					1
heavi						0.83%	1
here	0.24%						1
him		2.48%					1
howev							0
ideal		0.47%					1
ime	0.32%						1
import		0.47%					1
in		0.62%					1
insid	0.48%						1
intellectu							0
invent					12.31%		1
irrelev	0.32%						1
jhunix		0.47%					1

word	C1	C2	C3	C4	C5	C6	N
just	1.28%						1
kind							0
know		8.30%					1
known	0.48%						1
lamp					13.85%		1
launch						0.83%	1
law		0.70%					1
least		0.23%					1
left	0.72%						1
length		0.31%					1
less	0.96%						1
light		1.47%				0.41%	2
like		0.70%					1
line	1.61%						1
long	1.69%					0.21%	2
look	0.48%						1
mai							0
make		0.23%					1
mass	0.48%						1
matter	0.24%						1
mcgill						0.28%	1
me	0.72%	0.93%	5.56%				3
mean		0.23%					1
measur	1.61%	0.78%				0.41%	3
method		0.70%					1
microwav						1.10%	1
might		0.93%	5.56%				2
mind			5.56%				1
mine				6.15%			1
miner				9.23%			1
mint							0
mod			11.11%				1
model		0.78%					1
motion						1.03%	1

mous						0.34%	1
myer						1.38%	1
myself							0
never							0
new							0
nois		1.09%					1
note	0.96%						1
number		1.09%					1
object		0.39%					1
observ	2.65%	0.47%				0.21%	3
occur							0
origin						0.83%	1
our	1.28%						1
paradox	0.48%						1
paramet		0.47%					1
phenomena				7.41%			1
physic		2.33%					1
planet	2.65%						1
planetari	0.72%						1
pleas	0.24%						1
point	2.09%	0.70%					2
post		0.93%					1
predict		0.54%					1
problem		0.47%					1
process		0.47%					1
produc		0.47%					1
proof							0
proton		0.23%					1
prototyp		0.47%					1
prove							0
pseudo		0.47%					1
psi		0.62%		11.11%			2
publish				5.56%			1
question		1.16%					1

radiat						1.03%	1
random		1.09%					1
rate		0.47%					1
rather							0
reason		0.70%					1
redshift						1.10%	1
refer	0.48%						1
result	0.64%	0.39%					2
rider	6.02%					0.21%	2
rng		0.78%					1
roll		0.70%					1
room	0.96%						1
sai	0.40%	0.70%					2
said	0.40%	0.47%					2
same	1.69%						1
sampl		0.47%					1
saw		0.70%					1
scale						0.21%	1
sci		0.93%					1
scientif		0.31%					1
screen					10.77%		1
second	3.85%	0.47%					2
see	0.96%	0.39%				0.21%	3
seem							0
seriou						0.41%	1
shift						0.83%	1
signal		0.93%					1
simul		0.93%					1
simultan	2.89%						1
sinc	1.20%		5.56%				2
skeptic		1.16%					1
slower	0.96%						1
small						1.03%	1
solut		0.23%					1

somebodi		0.93%				1
someon			5.56%			1
someth		0.23%				1
space	1.28%					1
speed	0.96%	1.86%			0.28%	3
sr	0.40%					1
st			5.56%			1
standard		0.47%				1
stationari					0.41%	1
stephenson				9.23%		1
still	0.24%					1
stuart	0.64%					1
subject		1.16%				1
suggest	0.48%					1
suppress					14.44%	1
temperatur					0.62%	1
them	0.24%					1
theori		0.93%				1
thing	0.24%		7.41%			2
think	3.53%	0.70%		4.62%		3
though		0.31%				1
through				4.62%		1
tide	0.48%					1
time	3.93%	0.47%	7.41%		0.55%	4
timelin	2.57%					1
too						0
tool	1.28%					1
travel					0.41%	1
try	0.24%					1
turk		0.47%				1
two	0.96%					1
ulowel			11.11%			1
understand					1.03%	1
univers					0.21%	1

upon		0.70%					1
us	0.32%	1.40%				0.55%	3
uucp		1.16%					1
vadd						0.41%	1
vari		0.54%					1
variabl		0.31%					1
variat	0.48%	0.31%					2
veri		0.23%					1
view	0.72%						1
violat	0.32%						1
vision						0.28%	1
wai		0.47%					1
well	1.28%	0.70%					2
willner						0.83%	1
window		0.93%					1
work					4.62%		1
wrong	0.56%						1
wrt						0.96%	1
zdenek		3.65%					1
TRUE		0.62%					1

Appendix E Additional Charts

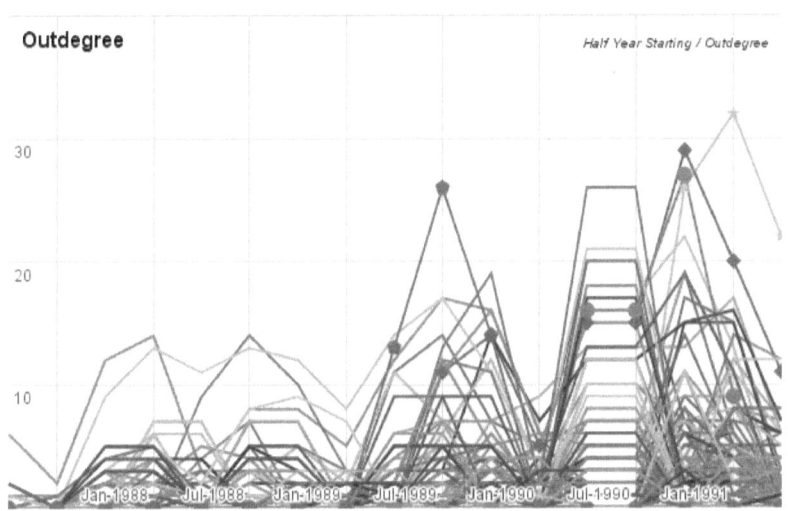

Figure 56 Outdegree for rec.birds; individuals with higher outdegree last for a long period of time

Figure 56 describes outdegree of individuals for rec.birds over a period of time. Individuals with higher values maintain their high positions.

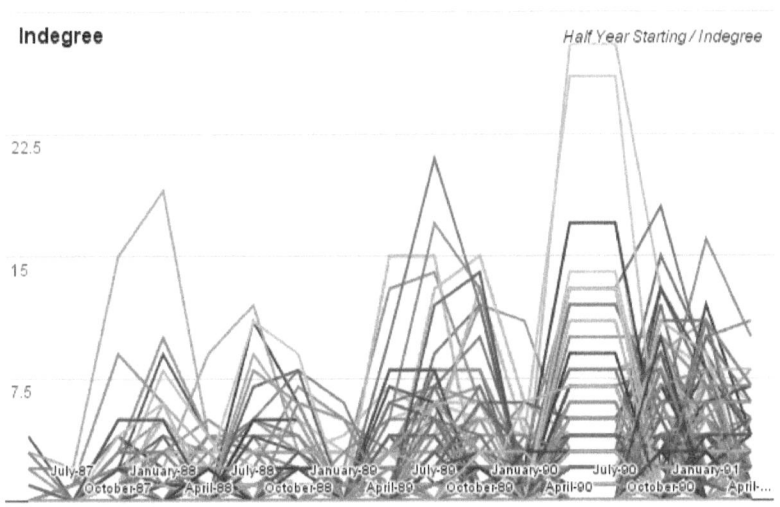

Figure 57 Indegree for rec.birds; people with high indegree last on top for a long period of time

Figure 57 describes the indegree of individuals in rec.birds over a period. Individuals who attain higher values maintain their positions.

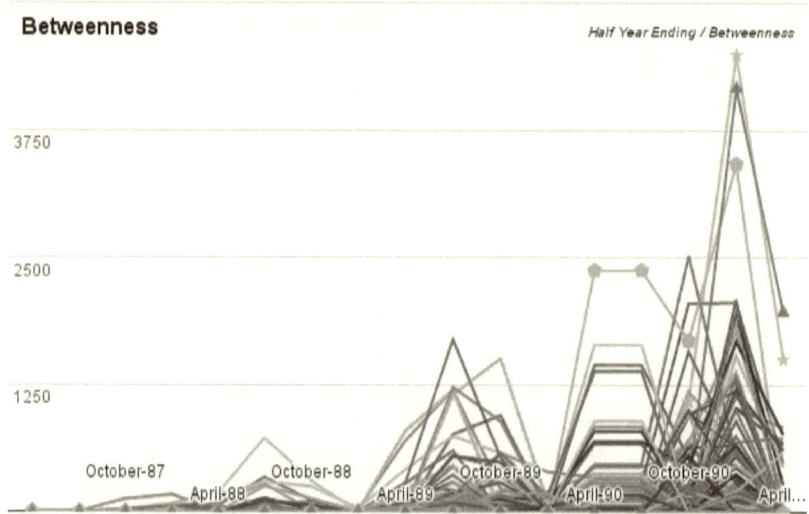

Figure 58 Betweenness for rec.birds; people with high betweenness stay at that level for long period of time

Figure 58 depicts the betweenness of individuals in rec.birds. This also shows a similar pattern,

users with higher values remain there.

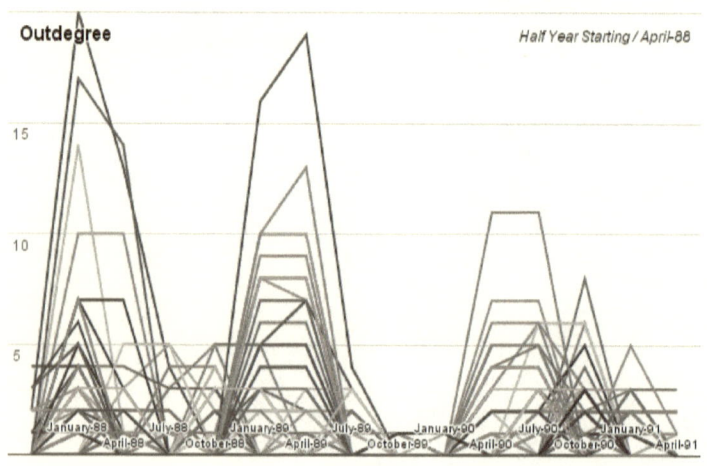

Figure 59 Outdegree for sci.psychology; people with high outdegree last for long periods of time

Figure 59 describes the outdegree of individuals in sci.psychology. The individuals who are at

high levels remain there.

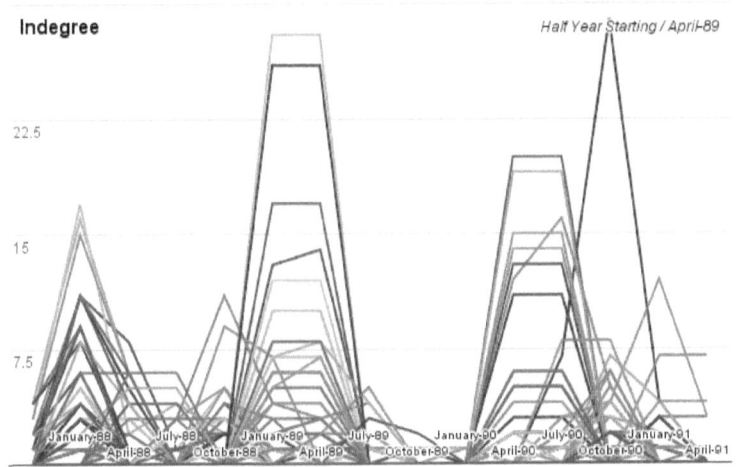

Figure 60 Indegree for sci.psychology; people with high indegree last for a long period of time

Figure 60 describes the indegree for sci.psychology, the individual who attains a position of primacy remains there.

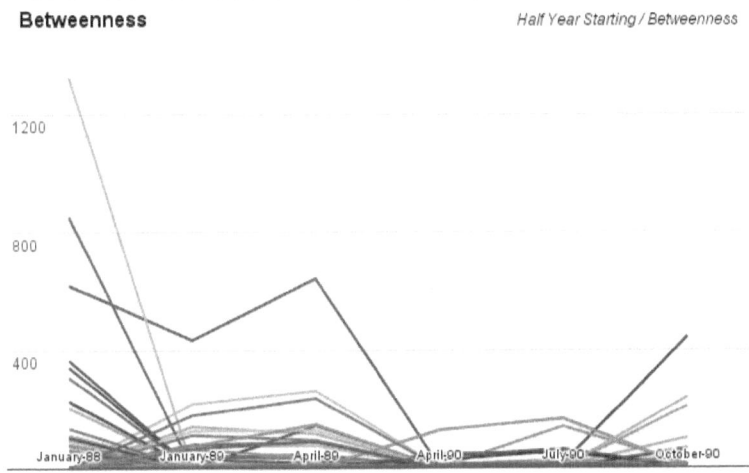

Figure 61 Betweenness for sci.psychology; people with high beweenness last for a long period of time

Figure 61 describes betweenness for the sci.psychology group, users who attain higher level remain at that level.

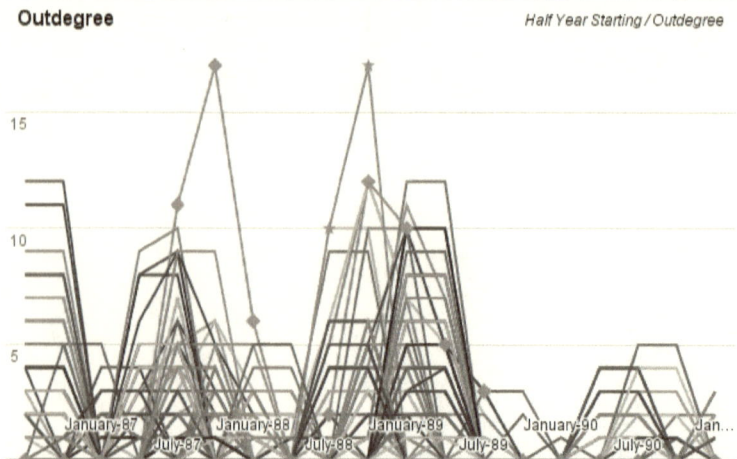

Figure 62 Outdegree of sci.lang; peopld with high outdegree last for a long period of time

Figure 62 describes the outdegree of the sci.lang group. The individual who attains higher outdegree remains at that level.

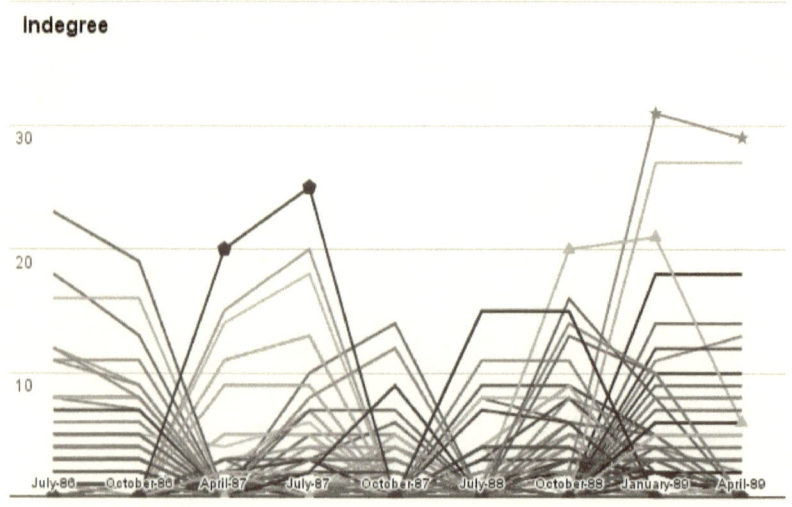

Figure 63 Indegree of sci.lang; people with high indegree last for a long period of time

Figure 63 describes the indegree of the sci.lang group. The individual who attains higher indegree remains at that level.

180

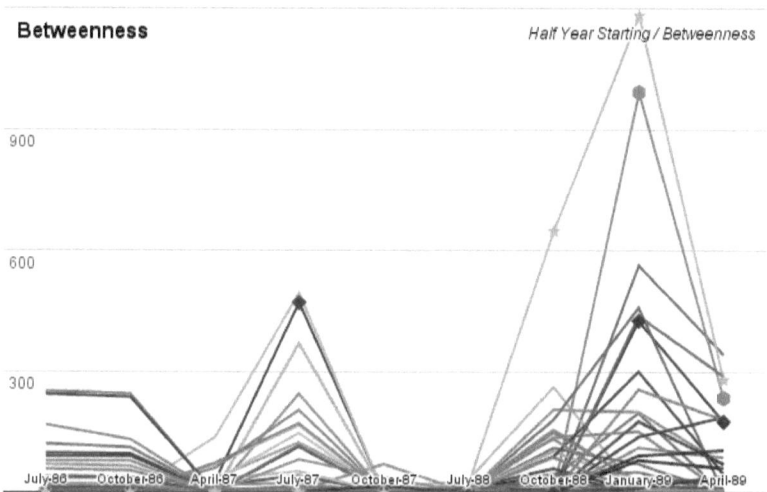

Figure 64 Betweenness of sci.lang; people with high betweenness last for a long period of time

Figure 64 describes the betweenness of sci.lang group. The individual who attains higher betweenness remains at that level.

Appendix F   SPSS Results

Descriptives

		N	Mean	Std. Deviation	Std. Error	95% Confidence Interval for Mean		Minimum	Maximum
						Lower Bound	Upper Bound		
contact_org_sme	No	34	1.50	2.164	.371	.75	2.25	0	6
	Yes	112	2.13	2.377	.225	1.69	2.58	0	7
	Total	146	1.99	2.337	.193	1.60	2.37	0	7
contact_sme_left	No	34	1.06	1.722	.295	.46	1.66	0	5
	Yes	112	1.28	1.645	.155	.97	1.58	0	6
	Total	146	1.23	1.660	.137	.95	1.50	0	6

Test of Homogeneity of Variances

	Levene Statistic	df1	df2	Sig.
contact_org_sme	3.460	1	144	.065
contact_sme_left	.008	1	144	.928

ANOVA

		Sum of Squares	df	Mean Square	F	Sig.
contact_org_sme	Between Groups	10.482	1	10.482	1.931	.167
	Within Groups	781.491	144	5.427		
	Total	791.973	145			
contact_sme_left	Between Groups	1.239	1	1.239	.448	.504
	Within Groups	398.302	144	2.766		
	Total	399.541	145			

Descriptives

		N	Mean	Std. Deviation	Std. Error	95% Confidence Interval for Mean		Minimum	Maximum	Between-Component Variance
						Lower Bound	Upper Bound			
contact_org_sme	No	110	1.23	2.079	.198	.83	1.62	0	6	
	Yes	36	4.31	1.348	.225	3.85	4.76	1	7	
	Total	146	1.99	2.337	.193	1.60	2.37	0	7	
	Model Fixed Effects			1.927	.160	1.67	2.30			
	Random Effects				1.720	-19.87	23.85			4.669
contact_sme_left	No	110	.70	1.365	.130	.44	.96	0	5	
	Yes	36	2.83	1.444	.241	2.34	3.32	1	6	
	Total	146	1.23	1.660	.137	.95	1.50	0	6	
	Model Fixed Effects			1.385	.115	1.00	1.45			
	Random Effects				1.192	-13.92	16.37			2.240

ANOVA

		Sum of Squares	df	Mean Square	F	Sig.
contact_org_sme	Between Groups	257.016	1	257.016	69.184	.000
	Within Groups	534.957	144	3.715		
	Total	791.973	145			
contact_sme_left	Between Groups	123.441	1	123.441	64.381	.000
	Within Groups	276.100	144	1.917		
	Total	399.541	145			

Test of Homogeneity of Variances

	Levene Statistic	df1	df2	Sig.
contact_org_sme	13.530	1	144	.000
contact_sme_left	1.096	1	144	.297

Descriptives

		N	Mean	Std. Deviation	Std. Error	95% Confidence Interval for Mean		Minimum	Maximum	Between-Component Variance
						Lower Bound	Upper Bound			
contact_org_sme	No	118	1.43	2.170	.200	1.04	1.83	0	7	
	Yes	28	4.32	1.389	.263	3.78	4.86	1	7	
	Total	146	1.99	2.337	.193	1.60	2.37	0	7	
	Model Fixed Effects			2.046	.169	1.65	2.32			
	Random Effects				1.687	-19.44	23.42			4.081
contact_sme_left	No	118	.83	1.458	.134	.56	1.10	0	5	
	Yes	28	2.89	1.423	.269	2.34	3.44	1	6	
	Total	146	1.23	1.660	.137	.95	1.50	0	6	
	Model Fixed Effects			1.451	.120	.99	1.46			
	Random Effects				1.204	-14.07	16.52			2.080

Test of Homogeneity of Variances

	Levene Statistic	df1	df2	Sig.
contact_org_sme	18.355	1	144	.000
contact_sme_left	.000	1	144	.983

ANOVA

		Sum of Squares	df	Mean Square	F	Sig.
contact_org_sme	Between Groups	188.908	1	188.908	45.107	.000
	Within Groups	603.065	144	4.188		
	Total	791.973	145			
contact_sme_left	Between Groups	96.252	1	96.252	45.700	.000
	Within Groups	303.289	144	2.106		
	Total	399.541	145			

Correlations

				contact_or g_sme	contact_s me_left	how_often _others_ne	when_othe rs_request	others_req uest_your_	peers_nee d_sys_pro
Kendall's tau_b	contact_org_s me	contact_or g_sme	Correlation Coefficient	1.000	.813**	.473**	.349**	.377**	.712**
			Sig. (2-tailed)	.	.000	.000	.000	.000	.000
			N	146	146	146	146	146	146
	contact_sme_l eft	contact_sme _left	Correlation Coefficient	.813**	1.000	.495**	.325**	.343**	.652**
			Sig. (2-tailed)	.000	.	.000	.000	.000	.000
			N	146	146	146	146	146	146
	how_often_oth ers_need_your _sme		Correlation Coefficient	.473**	.495**	1.000	-.169*	-.179*	.462**
			Sig. (2-tailed)	.000	.000	.	.031	.022	.000
			N	146	146	146	146	146	146
	when_others_r equest_help_y our_team_me mbers_can_hel		Correlation Coefficient	.349**	.325**	-.169*	1.000	.920**	.315**
			Sig. (2-tailed)	.000	.000	.031	.	.000	.000
			N	146	146	146	146	146	146
	others_request _your_help_for _sys_org_proc		Correlation Coefficient	.377**	.343**	-.179*	.920**	1.000	.356**
			Sig. (2-tailed)	.000	.000	.022	.000	.	.000
			N	146	146	146	146	146	146
	peers_ne ed_s ys_p		Correlation Coefficient	.712**	.652**	.462**	.315**	.356**	1.000

185

			Sig. (2-tailed)	.000	.000	.000	.000	.000	.
			N	146	146	146	146	146	146
Spearman's rho	contact_org_s me	Correlation Coefficient	1.000	.924**	.523**	.383**	.417**	.803**
		Sig. (2-tailed)	.	.000	.000	.000	.000	.000
		N	146	146	146	146	146	146
	contact_sme_l eft	Correlation Coefficient	.924**	1.000	.551**	.368**	.382**	.758**
		Sig. (2-tailed)	.000	.	.000	.000	.000	.000
		N	146	146	146	146	146	146
	how_often_oth ers_need_your _sme	Correlation Coefficient	.523**	.551**	1.000	-.179*	-.191*	.494**
		Sig. (2-tailed)	.000	.000	.	.031	.021	.000
		N	146	146	146	146	146	146
	when_others_r equest_help_y our_team_me mbers_can_he	Correlation Coefficient	.383**	.368**	-.179*	1.000	.943**	.342**
		Sig. (2-tailed)	.000	.000	.031	.	.000	.000
		N	146	146	146	146	146	146
	others_request _your_help_for _sys_org_proc	Correlation Coefficient	.417**	.382**	-.191*	.943**	1.000	.389**
		Sig. (2-tailed)	.000	.000	.021	.000	.	.000
		N	146	146	146	146	146	146
	peers_need_s ys_proc_help	Correlation Coefficient	.803**	.758**	.494**	.342**	.389**	1.000
		Sig. (2-tailed)	.000	.000	.000	.000	.000	.
		N	146	146	146	146	146	146

**. Correlation is significant at the 0.01 level (2-tailed).

*. Correlation is significant at the 0.05 level (2-tailed).

Appendix G　　Publications

Journal Publications

Avasthi, Vinay, & Dey, S. (2015). Loss in tacit knowledge because of employees attrition Vinay Avasthi Shubhamoy Dey. International Journal of Intercultural Information Management\, 5(x), 1–18. https://doi.org/10.1504/IJIIM.2015.072541

Conference Proceedings

Avasthi, V., Dey, S., Jain, K. K., & Mishra, R. (2015). The Evolution of Knowledge in Communities of Practice. Proceedings of the 2015 Conference on Research in Adaptive and Convergent Systems, 1–6. https://doi.org/10.1145/2811411.2811528

Avasthi, V., Dey, S., Venkatagiri, S., Jain, K. K., & Mishra, R. (2015). Knowledge Networks and Knowledge Adjacencies. In Proceedings of the 8th Annual ACM India Conference (pp. 111–116).

www.ingramcontent.com/pod-product-compliance
Lightning Source LLC
Chambersburg PA
CBHW030630220526
45463CB00004B/1475